C0-CCT-534

No. 1932
$17.95

AM STEREO & TV STEREO
NEW SOUND DIMENSIONS

STAN PRENTISS

Other TAB Books by the Author

No. 1532 *The Complete Book of Oscilloscopes*
No. 1682 *Introducing Cellular Communications: The New Mobile Telephone System*
No. 1632 *Satellite Communications*

TV SIDEKICK is a trademark of Modulation Sciences Inc.
C-QUAM and QUAM are trademarks of the Motorola Corp.
DYANAMIC NOISE REDUCTION is a registered trademark of the National Semiconductor Corp.
WALKMAN is a trademark of the Sony Corp.

FIRST EDITION

FIRST PRINTING

Copyright © 1985 by TAB BOOKS Inc.
Printed in the United States of America

Reproduction or publication of the content in any manner, without express permission of the publisher, is prohibited. No liability is assumed with respect to the use of the information herein.

Library of Congress Cataloging in Publication Data

Prentiss, Stan.
AM stereo and TV stereo—new sound dimensions.

Includes index.
1. Stereophonic sound systems. I. Title. II. Title:
A.M. stereo and T.V. stereo—new sound dimensions.
TK7881.8.P74 1985 621.38'0413 84-23982
ISBN 0-8306-0932-6
ISBN 0-8306-1932-1 (pbk.)

Cover photograph courtesy of Delco Electronics Division, General Motors Corporation.

Contents

Acknowledgments

I wish to acknowledge with more than considerable gratitude the very helpful aids and suggestions of largely engineering personnel from EIA, the federal government, and various affected industries:

Ed Tingley and Thomas Mock, EIA; John W. Reiser, Bruce Franca, Ralph Haller, Jim McNally, John Reed, FCC; Lee Hoke, Jr., Pat Wilson, Ray Guichard, Walt Schwartz, NAP Consumer Products; Leonard Kahn, Kahn Communications; Charles Smaltz, National Semiconductor; Dr. Jim Carnes, Mike Deiss, RCA; Peter Foldes, Ed Polcen, Bob Hansen, Jim White, Zenith; Hratch Aris, Panasonic; David Lynch, Sanjar Ghaem, General Electric; Mike Rau, Ed. Williams, NAB; Eric Small, Modulation Sciences; Ray McMartin, McMartin Industries; Ed Onders, Kahn/Hazeltine; Lew Eads, Delco Radio; R.R. Weirather, Harris Corp.; Vernon Collins, Continental Electronics; Larry Ecklund, Dick Harasek, Chris Payne, Motorola; Wendell Bailey, National Cable TV Assn.; Jim Allen, Ford; Stan Saler, Broadcast Electronics; Bob Orban, Orban Associates; David Kawakami, Neil Cunningham, John Strom, Sony: Stanley Salek, Broadcast Electronics; Bud McDowell, Magnavox; and especially Pieter Fockens of Zenith, without whom much of the hard technical information on TV stereo sound could not have been accurately developed. Additional special thanks are due to Frank Hilbert of Motorola, our devout critic, who helped clarify and correct a number of technical points in AM stereo.

Introduction

If you're looking for a double-barreled bargain, this should be one! Amplitude Modulation (AM) *radio* stereo and Frequency Modulation (FM) *television* stereo are fully characterized and explained in a single publication that includes thorough subject research, regulatory and electrical FCC evaluation of both systems, and the latest available information on transmitters and receivers now at broadcasters and in the field.

You'll read about extensive testing and recommendations by the Electronic Industries Association, the National Cable Television Association, the Federal Communications Commission, and many others involved over the years. Then you will be able to develop an in-depth understanding of how one system applicant and not another was chosen as the American standard. You'll also see why the FCC first approved Magnavox's AM stereo proposal and then rejected it in the face of more than considerable industry unhappiness, along with the resulting delay of almost four years in general AM stereo broadcast/receiver availability. Conversely, a prompt/one-system FCC approval of Zenith-dbx in March 1984 brings stereo receivers to the market in only a few months, even though substantial developments in TV transmitter stereo exciters won't be available until later this year.

Establishing major electrical products as highly desirable items for many or most consumers becomes an extremely difficult (and often hazardous) task as evidenced by the recent rise and fall of the video disc. But AM and FM stereophonic and bilingual sound seem to be more probable necessities than some of their past competition. If so, both services and their technically advanced products will soon occupy a very select commercial slot in the United States. In the meantime, you have all the up-to-date informatin that's available for print just now, including some industry suggestions on how to modify your own AM monophonic radio to receive stereo, which is often impractical.

Chapter 1

AM Stereo Initial Systems and Tests

Much of this chapter is based on AM stereo systems first proposed and tested by NAMSRC, the National AM Stereophonic Radio Committee, formed September 24, 1975, and jointly sponsored by The Electronic Industries Association, The National Association of Broadcasters, The National Radio Broadcasters Association, and the Broadcasting, Cable and Consumer Electronics Society of the Institute of Electrical and Electronics Engineers, (EIA, NAB, NRBA, and BCCE, respectively). As a result of these studies and tests, considerable system modifications and revisions have since been undertaken by the various manufacturers, and one of the original proponents has, apparently, not pursued his efforts further. Whether he will reappear at some future date is not known.

In later chapters, descriptions of all systems will be covered, including whatever modifications have been made and are available for publication. For the moment, we'll concentrate on transmitters, but receivers and ICs are to receive their share of attention too, since many will be introduced to the market in the 1984-1986 time frame. Eventually, we would expect that AM stereo will easily rival FM stereo in popularity, and may even exceed it on the road because of greater distance reception and absence of strongest-station capture effect so prevalent in all FM systems.

At the moment, and probably for some time to come, fierce competition between the three major remaining systems (of the five originals) continues among broadcasters and transmitter exciter makers, so many receiver manufacturers must also choose to go with a single system, or try and wait out a possible multisystem decoder suitable for the entire country. Although improvements to both transmitters and receivers will continue, as in television, many systems are adequate right now and users will get their money's worth with whatever they choose regardless of the ultimate system.

Consequently, what we propose in this introductory chapter is to introduce you to the initial systems as proposed and partially tested, proceed with the Federal Communications Commission's Decisions in the next chapter, and finally wind up with the latest developments in both transmitters and available receivers. Such a "step" approach should supply most of the relevant details and Federal actions without painful plunges into the icy waters of advanced technology before due preparation and gentle immersion.

As we proceed, you'll note that Delco (GM), Ford, and Chrysler will be offering mobile products (all C-QUAM Motorola), while portables and home units come from such places as Korea, Taiwan, Hong Kong, and Japan. Four-way decoders are being worked on by Sansui and Hazeltine, while Sony has two portables already in the marketplace. True, it's easier and better engineering to design a single system transmitter and receiver, but what the ultimate resolution will be in the years to come is unpredictable at the moment since the retail selling price will be a dominant factor.

Consequently, after being somewhat strongly admonished, we'll try to stick to facts rather then over enthusiastic predictions. By so doing, AM stereo may get a better lease on life and a faster start than loose stargazings of wonders to come. It would be good for the industry, however, if more expensive, but good receivers were developed in time to receive quality reception on *all* the remaining systems. A two-chip approach, we're told, is possible but certainly costly.

Therefore, with the foregoing in mind, let's begin a review of AM stereo beginnings and proceed, chapter by chapter, with what's available today. Systems, as you will see, do change, so don't be confused by the tentative beginnings and the final products; they may be considerably different.

BASIC SYSTEMS

The five initial systems, only three of which were tested by NAMSRC, are Magnavox, RCA-Belar, Kahn, Motorola, and Harris. The Belar system is the one not active at this time (1985). All of these transmitting arrangements had one thing in common: *to reproduce standard amplitude modulation for radio compatibility, the L+R had to be conventional AM*. The big trick in stereocasting was, and is, to combine L+R and L−R into one envelope, producing virtually full 15 kHz bandpass without undue distortion from the subtractive channel. Different approaches produced different results, as we shall see.

Magnavox

This is best described as an amplitude and phase (AM/PM) modulation system in which L−R becomes linearly phase-modulated with a peak phase deviation of 1 radian. A stereo 5 Hz subaudible tone frequency-modulates the carrier with a deviation of some ±20 Hz for stereo identification. This tone would also be used, according to Magnavox, to transmit some very slow digital information such as transmitter identification or text.

Transmission. AM stereo is broadcast by disabling the standard AM broadcast transmit oscillator and then driving the transmitter with AM stereo rf. The 5 Hz stereo identification with some 4 radians of total phase deviation combines with a phase modulator, adding L−R to the carrier at a low level stage in the broadcast transmitter chain. Both L+R and L−R are then modulated on available rf, with delay circuits equalizing the time delays or advances between the two. As you can see in the simplified block diagram (Fig. 1-1), phase modulation for the stereo portion enters the phase modulator following the carrier generator via the audio matrix, while L+R continues through the audio matrix and into the amplitude modulator for L−R/L+R combining in the final rf power amplifier.

AM transmitters are converted to stereo use with the inclusion of an exciter. This unit receives ac power, the audio oscillator test tone—when the system is being tested—stereo left and right signals, a synthesized phase reference, and phase modulation. It delivers rf drive and modulated stereo to the broadcast transmitter.

Audio matrix and low level rf drive amplifiers are apparent in the exciter (Fig. 1-2), as well as stereo metering for program content. There's also an audio delay circuit that is able to produce accurate time delays from 0 to 77.5 μsec in 2.5 μsec steps. Audio delays may be inserted in either L+R or L−R information to compensate for any time restraints in either channel. Afterwards, L+R is both amplified and transformer-isolated for signal

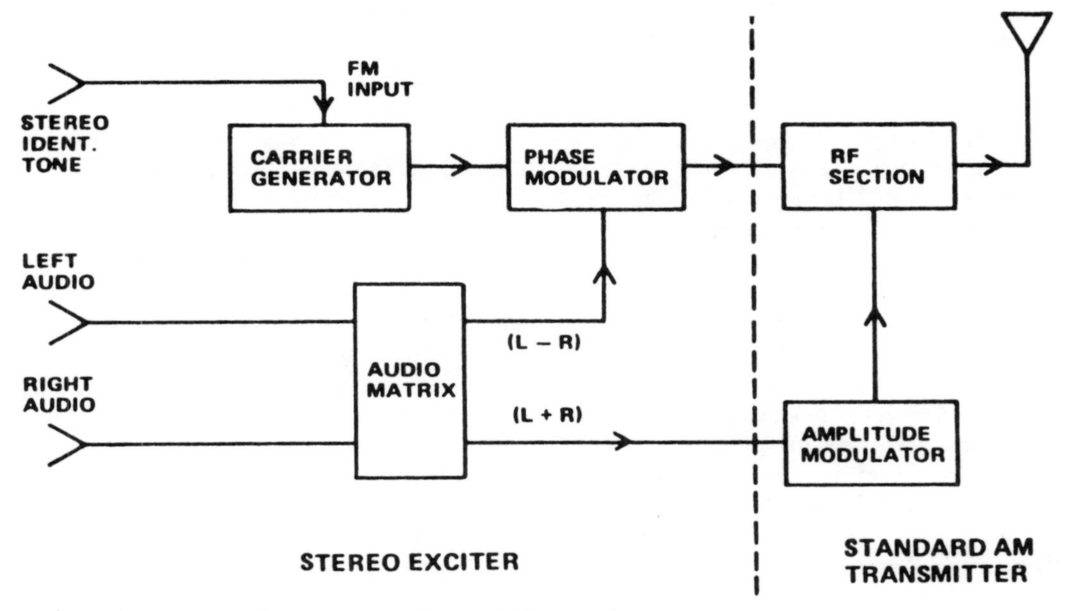

Fig. 1-1. Basic block diagram of stereo system (courtesy Magnavox).

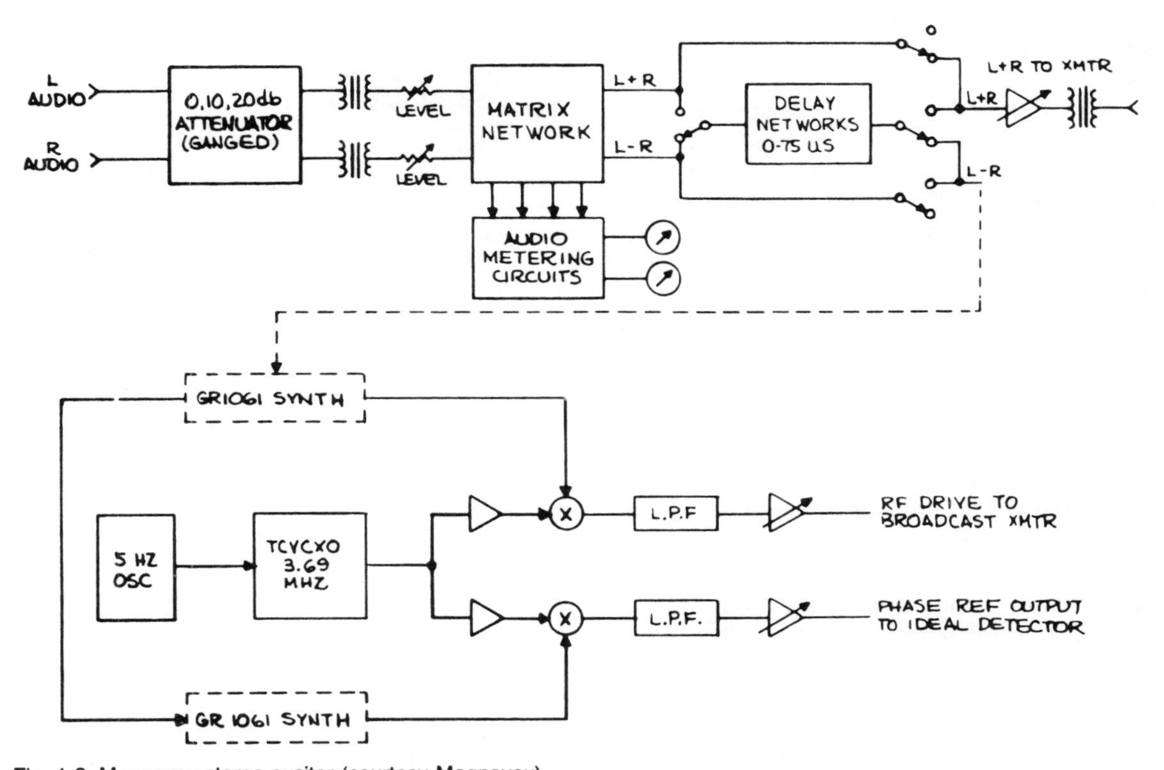

Fig. 1-2. Magnavox stereo exciter (courtesy Magnavox).

the existing AM transmitter. As the diagram illustrates, a 5 Hz oscillator generates a tone signal for a special receiver stereo indicator lamp, modulated on the 3.69 MHz crystal-controlled oscillator.

Two outputs pass through low pass filters to the rf drivers and phase reference outputs. Being above the 0.54 to 1.6 MHz range of the authorized broadcast band, this oscillator is heterodyned down to suitable transmitter carrier frequencies; the two synthesizers amplifying and upbeating the 5 Hz modulating signal to broadcast frequencies along with phase-modulated L−R audio. Built-in test equipment will measure both negative AM peak modulation and low level modulation on both AM and phase modulated signals from the transmitter, enabling verification of both crosstalk levels between the AM and PM, as well as apparent residual noise levels.

Magnavox Receivers. Receivers for this type of AM stereo generation detect separate amplitude and phase signal variations, then a matrix produces the usual left (L) and right (R) audio outputs. The receiver block diagram is shown in Fig. 1-3.

Received signals are radio frequency and intermediate frequency amplified, bandlimited, and processed according to customary procedures and standard high/low automatic gain (AGC) control. IF amplifiers contain a dual bandwidth filter system, a dual AGC loop, plus signal level and center tuning meters designed for user convenience and precise station selection. IF filters offer sharp skirt rolloffs and low passband ripple with "normal" bandwidths of 6 kHz and "fidelity" bandwidths of 12 kHz. These maximize both groundwave and skywave reception under most signal conditions. Selection is by a Normal/Fidelity front panel switch, with 455 kHz IF information passing through either a pair of 4-pole lumped constant filters in series or one with 8 poles, followed by an rf AGC and tuning meter detector and IF amplifier. Succeeding the IF amplifier is the second portion of the rf AGC detector as well as signal inputs to the center tune drive, full wave L+R envelope detector and limiter for the L−R phase detector.

An MC1355 IC limiter (not shown) removes L−R envelope modulation for the phase detector which begins the process of phase-locked loop detection, aided by a loop filter and amplifier, a voltage-controlled oscillator, and high pass noise detector. The 5 Hz signal lamp subsonic signal is then passed through a low pass filter, detected, and then on to part of the front panel stereo indicator control logic. The other half picks up AGC current to the tuning meter, via a level detector, and, together, they turn on the front panel stereo lamp whenever there is incoming AM stereo.

Outputs of the loop filter go not only to the 5 Hz low pass filter but also to the voltage-controlled oscillator and high pass noise detector which triggers control logic for the mute switch. The purpose of this switch is to render tuning transients silent, but may be defeated by an operator-controlled toggle switch.

After a 16 μsec AM signal delay, L+R (mono) and L−R (stereo) are once more joined in a matrix, producing pure left (L) and right (R) signals for the 12 W stereo audio system and connected speakers or headphones.

Motorola's C-QUAM

Motorola has done a great deal of work on its C-QUAM system, both before and after the FCC's open market decision and, mobile-wise, it's apparently paying off handsomely, at least among first introductions. C-QUAM translates to a Compatible Quadrature Amplitude Modulation system that overcomes negative modulation problems of AM/PM approaches. But some say it's really an AM/PM arrangement where "phase modulation is increased on dips in the L+R modulation; and such distortion has to be removed in the receiver." Motorola archly replies "there's *no* distortion with proper decoding."

Motorola claims advantages of:

☐ An L−R performance on par with quadrature modulation that's free of AM noise characteristics during modulation.

☐ Minimum spectrum spread.

☐ Equal or better compatibility versus other systems tested.

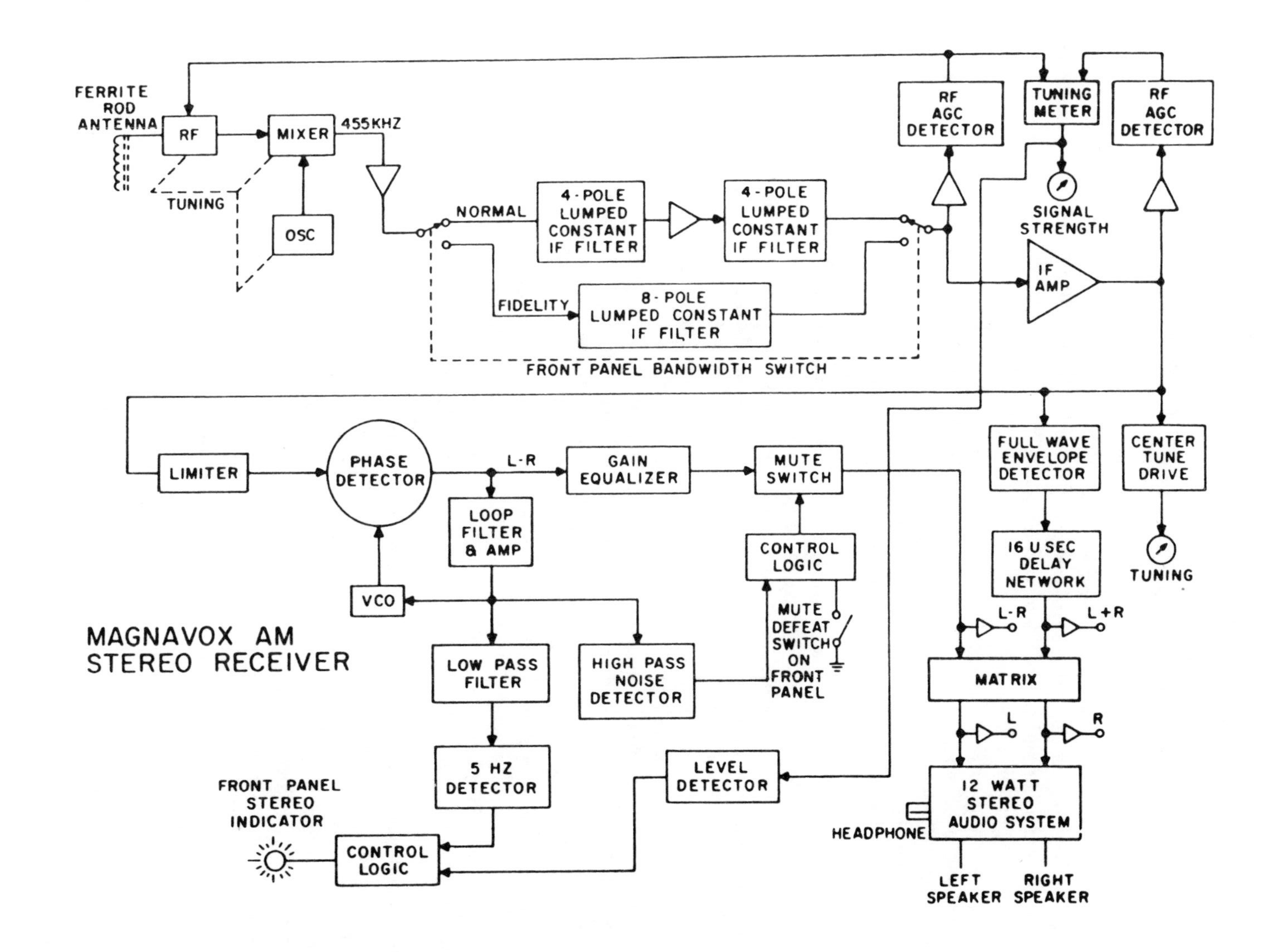

Fig. 1-3. Magnavox AM stereo receiver (courtesy Magnavox).

☐ Will permit various decoding methods.

☐ Includes a pilot tone that's independent of amplitude modulation.

Motorola compares its system to that of color television where I and Q signals are combined in carrier-suppressed modulation and then joined with luminance information on the main carrier. Matrixing of the AM radio signals, according to Motorola, "can be accomplished either after detection or in the process of detection (so-called rf matrixing) by choice of the angle of synchronous detection." Quadrature channel noise is said *not* to be a function of the in-phase channel.

Quadrature Transmission. As seen in Fig. 1-4, a single crystal-controlled carrier frequency generator supplies transmitter No. 1 and transmitter No. 2 with carrier energy in quadrature. Left and right audio then modulate their respective carriers, with power outputs of each transmitter summed and combined in a common antenna load.

The crystal oscillator first supplies rf drive to a balanced modulator producing a suppressed carrier with double sidebands L(t)+R(t) of amplitude modulation. As the carrier is phase shifted 90 degrees and applied to the second balanced modulator, L(t)−R(t) sidebands are generated in quadrature with the others. When the two are combined in a summer, the output is identical to the two transmitter outputs. By using a limiter, phase modulation may be added to the signal directly "at the transmitter exciter input." An envelope can then be derived from the composite signal by an envelope detector and used as a modulation audio input. According to the National AM Stereophonic Cmte., quadrature-modulated stereo has been extensively field tested around New York as early as the 1960s by both CBS and Philco using a transmitter very similar to the one described above.

NAMSC applauds the idea of simple modifications to any transmitter exciter being able to generate quadrature amplitude modulated signals (QUAM), and reception of the "sum signal" on existing receivers with very simple decoding. "It has well know advantages," they continue, "of spectrum efficiency, noise performance, and matrix decoding options."

But there is a problem, they say, in the standard monophonic receivers when transmitted signals contain a significant amount of stereo. The stereo receiver cleanly separates L+R and L−R information into left and right information, but the monophonic set has to reproduce equal, L=R, and the stereo outputs are identical. But when L is not equal to R (during stereo), the envelope detector does not produce a linear output and contains distortion.

C-QUAM. To produce receiver linearity, the QUAM output can be modified to permit linear detectors to accept a compatible sum signal. This has come to be known as C-QUAM and is accomplished in this way:

QUAM is first generated, limiting occurs, and is then remodulated with 1+L+R, according to NAMSC. As illustrated in Fig. 1-5, Lt+Rt undergo one process, while Lt−Rt and the pilot tone undergo another. Both are controlled by the transmitter crystal. The two resulting signals are summed and limited as shown, reaching the AM transmitter without the aid of an envelope detector, $\pi/2$, of course, indicating a 90-degree phase shift. Audio modulation is then applied directly to the AM transmitter which produced what we know as C-QUAM.

The Receiver. In the receiver, Fig. 1-5 indicates the difference between the monophonic and stereo reception. Envelope detectors may now demodulate linearly since their outputs correspond to transmitter modulations and are substantially compatible for right and left intelligence. In stereo reception, NAMSC says that "the C-QUAM signal is divided by the cosine of the angular modulation so that the original QUAM signal can be restored and detected by a pair of quadrature detectors. In an actual stereo set, you can have a carrier level modulator, quadrature phase detector for phase-locked loop generator, a vco feeding an in-phase detector, and a pair of 45-degree phase shifters into their respective balanced modulators. These produce both right and left outputs for final amplifiers and speakers. Such balanced modulators, of course, are actually synchronous detectors in the receiver. A 90-degree phase shifter after the vco supplies an

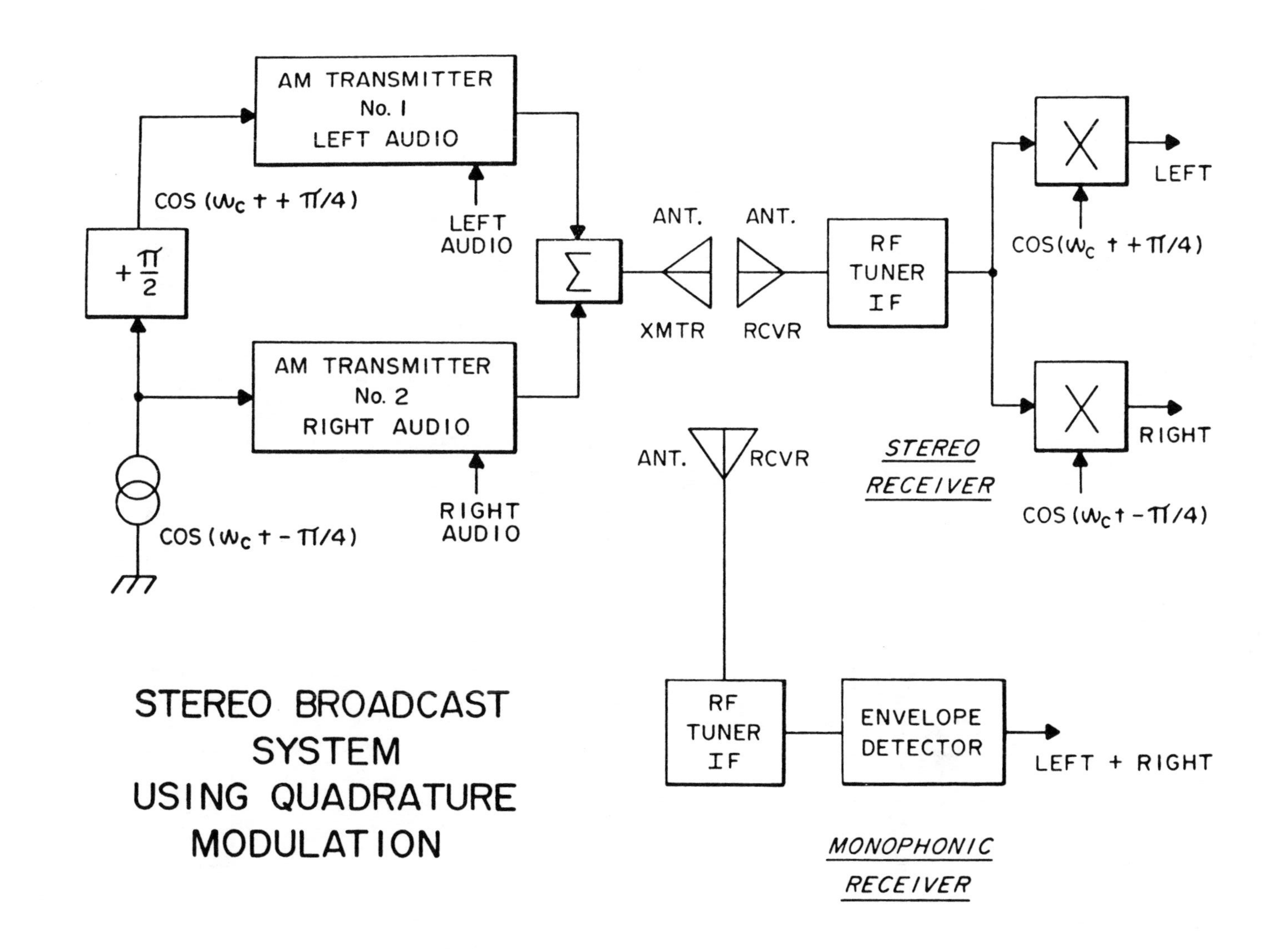

Fig. 1-4. The Motorola C-QUAM system (courtesy Motorola).

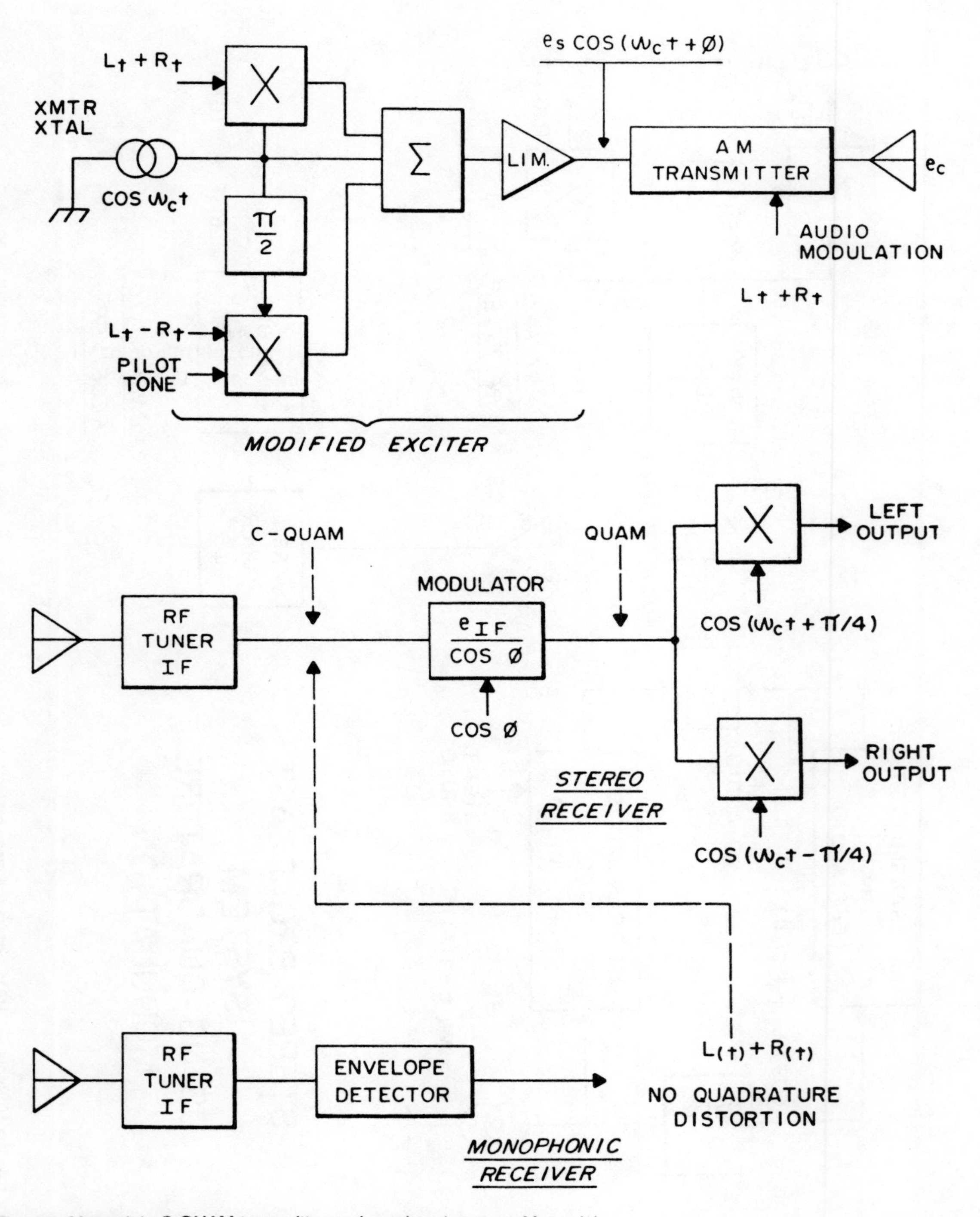

Fig. 1-5. Motorola's C-QUAM transmitter and receiver (courtesy Motorola).

in-phase feed to the IF carrier through its in-phase detector.

Because QUAM and/or phase modulation receivers usually have both in-phase and quadrature detectors for squelch and vco control, the only additional element needed for any C-QUAM receiver is a carrier level monitor.

C-QUAM to QUAM conversion, as suggested by the Committee, could take the form of differential amplifier input from the IF amplifier (Fig. 1-6). The in-phase detector supplies a current source through diodes D1 and D2 to the bases of the differential amplifier, one input of which comes from the IF amplifier, with the other base biased by a resistor to ground. Both T1 and T2 then operate as differential signal sources (one conducts when the other doesn't). With the IF amplifier output reaching the base of T1 with drive from D2, the differential amplifier responds accordingly to its inputs, supplying drive current to left and right switching gates "controlled by the demodulating reference signal." NAMSC says that such current sources are already available in a QUAM receiver and only the two diodes, base resistors, and the $i_0 \cos \phi$ are the added elements. C-QUAM, however, may be decoded by a number of methods, and we'll undoubtedly see changes when the final version is described in a later chapter.

The *C-QUAM encoder* is shown in Fig. 1-7. Left and right audio inputs pass through 600-ohm balanced transformers and their attached potentiometers, which are gain controls delivering sound to variable delay lines as well as the several matrices.

The upper L+R matrix is the amplitude circuit supplying both the initial power amplifier and also

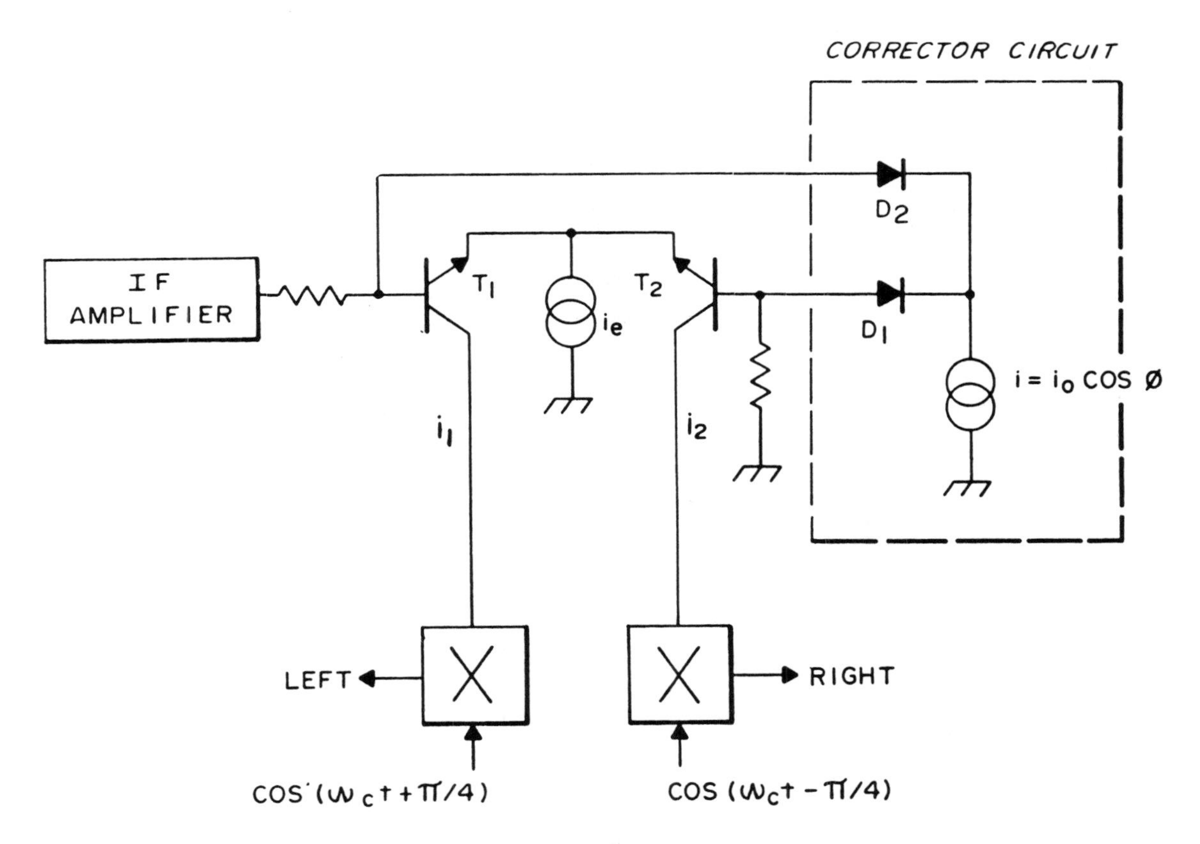

Fig. 1-6. Analog modulator or divider (courtesy Motorola).

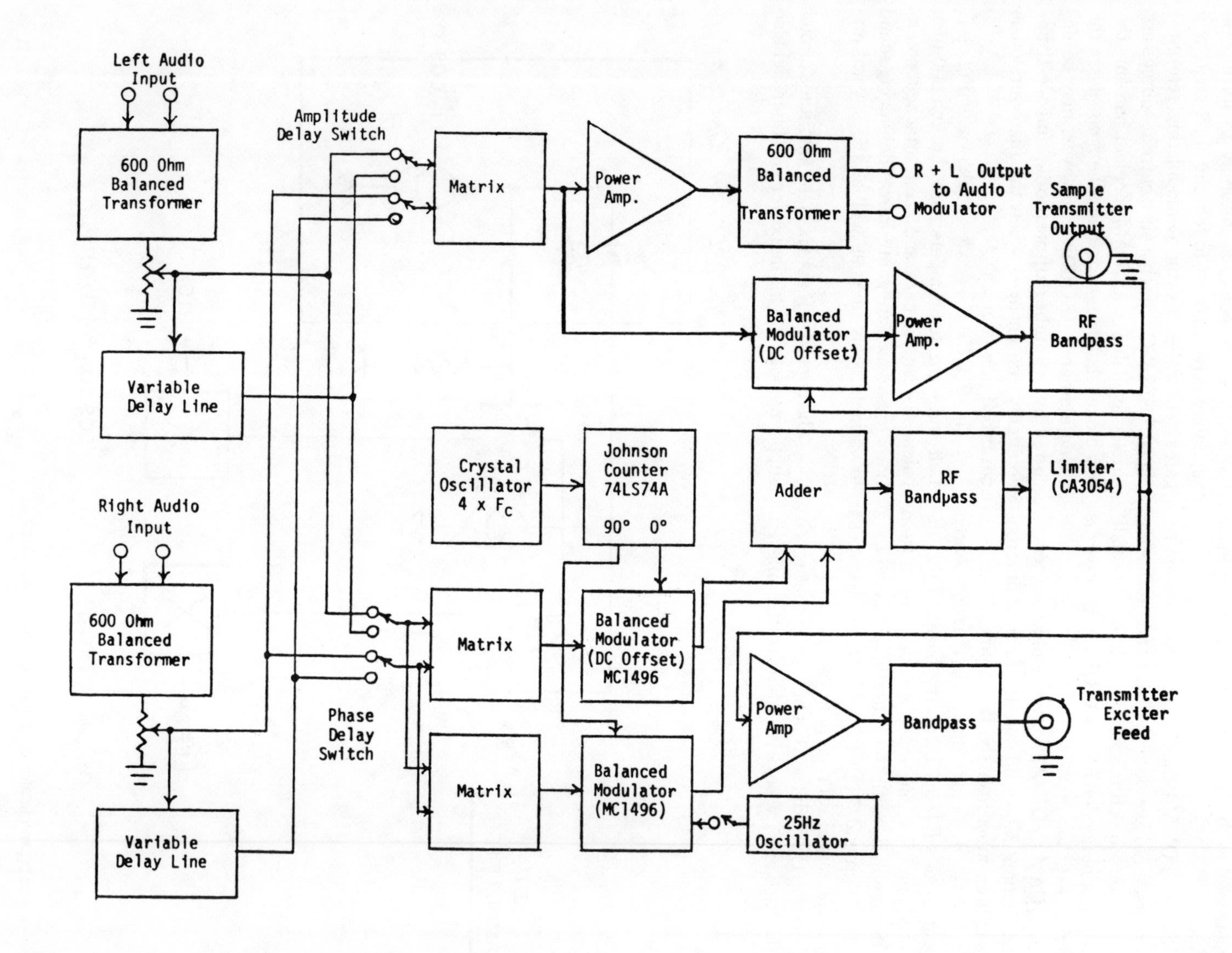

Fig. 1-7. Motorola's C-QUAM transmitter encoder (courtesy Motorola).

the balanced modulator (with dc offset). The lower L+R and L−R pair are the phase matrices, and both sets may be switched to direct or delayed audio to compensate for any group delay differences between rf and modulation that may arise.

Outputs of the two lower matrices continue into a pair of balanced modulators, L−R being phase shifted 90 degrees by a Johnson counter whose crystal-controlled frequency is 4 times that of the carrier frequency (F_c). A 25 Hz oscillator also enters the L−R balanced modulator at 5 percent modulation through a switch supplying the pilot tone.

L+R and L−R information are now added, routed into rf bandpass and limiter and on to a final power amplifier and bandpass output to the transmitter exciter. Another portion of this rf bandpass signal also goes to the balanced modulator (with dc offset), into a power amplifier and another rf bandpass which, according to NAMSC, acts as a "small transmitter"—apparently an auxiliary of some description, probably used in testing. In the receiver of course, envelope and quadrature detectors pick up and process incoming rf, with another Johnson counter doing the quadrature phase shifting. You'll also have the usual voltage-controlled crystal oscillator, 25 Hz pilot tone detector, and the left and right matrix.

The Belar System

First extensively tested by RCA in 1959, Belar Electronics Laboratory, Inc., has offered an AM-FM stereo system where right and left audio enters a matrix to produce L+R and L−R, with the L−R component being pre-emphasized to frequency modulate rf transmitter drive. These modulation constants have a low frequency deviation of 320 Hz (originally 1.2 kHz), and a pre-emphasis time constant (tc) of 100 microseconds (μsec). With adequate IF filtering and FM detector response, both FM and AM portions of the signal can be independently detected to produce L+R and L−R. Afterwards, linear addition and subtraction will recover both right and left audio channels. A basic block diagram is shown in Fig. 1-8 of both the transmitter and receiver.

Left and right sound signals enter an audio matrix which results in L+R and L−R outputs. L+R information goes directly to the AM modulator, while L−R is first pre-emphasized and then applied to an FM modulator which is also receiving excitation from an rf source. This FM information then passes directly into the existing AM transmitter where it is combined with L+R standard AM modulation and broadcast.

Rf and IF circuits are fairly standard for better quality AM receivers. After the IFs, however, AM information proceeds to an envelope detector, while FM is limited, removing the AM component, and then demodulated by the well-known frequency discriminator. After this, FM becomes de-emphasized to restore original amplitude and improve signal-to-noise (S/N) of L−R intelligence. Both signals are then simply audio-matrixed in a linear addition-and-subtraction process to recover left and right speech or music sound—a method sometimes dubbed "simulcast" when separate FM and AM receivers are used for stereo broadcast listening.

AM-FM Stereo Exciter. A more detailed block diagram of the exciter as tested by NAMSRC at radio stations WGMS, WTOP, and WBT is illustrated in Fig. 1-9. A 3 MHz, crystal-controlled oscillator, divided down by 300 to 10 kHz, feeds a phase comparator, loop filter, and voltage-controlled oscillator, with a divide by N frequency synthesizer, delivering a range of selected frequencies between 500 and 2,000 kHz. Loop constants are carefully designed so the vco may be linearly frequency modulated without what NAMSRC terms *loop following,* which probably means undesirable loop oscillation and control effects between input and output. Modulated FM is now amplified for sufficient transmitter drive and rf takes the place of an ordinary crystal oscillator in the AM transmitter.

FM stereo modulation of the vco comes from the left input via a balun (balanced-to-unbalanced transformer), a left input level control and L/R matrix, including a 100 μsec pre-emphasis. The L+R matrix output is subject to another level potentiometer before reaching a negative peak limiter which will hold or clamp negative modulation peaks

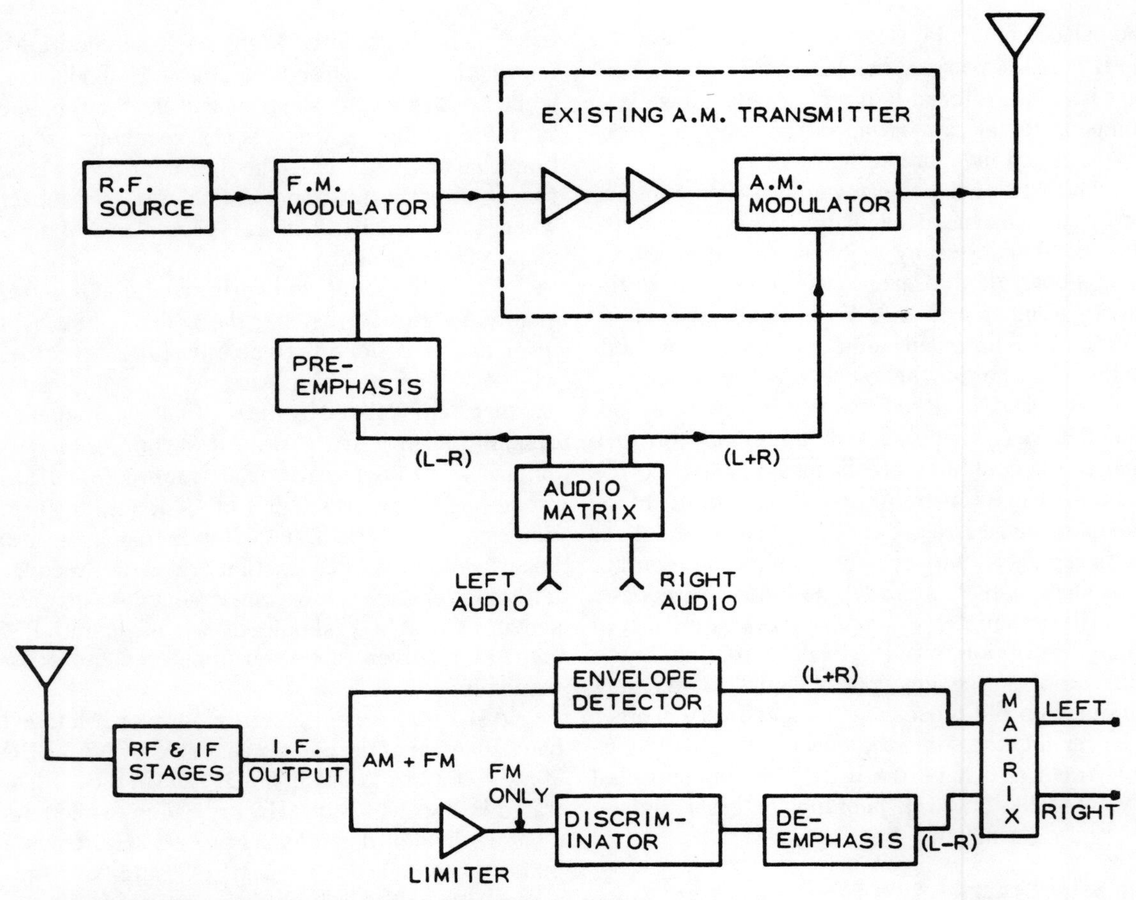

Fig. 1-8. Basic transmitter receiver for Belar AM-FM stereo system (courtesy Belar).

anywhere between 90 and 100 percent, according to selection. Positive peak excursions were unlimited in the NAMSRC tests. L+R audio information then continues on to the transmitter via a 600-ohm balanced transformer.

Rf along with L−R and L+R information may now be broadcast on the station's AM transmitter and received by either standard or specifically equipped radios designed for stereo reproduction.

The Receiver. A stereo station monitor used by NAMSRC in these tests was supplied by Belar, which included a carrier level monitor, FM discriminator, positive and negative modulation meters, right and left channel metering, and the required 100 μsec de-emphasis circuits. The receiver, itself, however, seems more important to most readers than meter movements, so we'll talk about that in Fig. 1-10.

This is a very basic block diagram of the receiver, but most significant points have been previously discussed. A signal path with brief descriptions is all that should be required to complete the Belar portion of our discussion.

Electromagnetic energy from broadcasts reaches the shielded and preamplified loop antenna, generating a signal that is routed to the receiver's rf circuits. One 2 MHz low pass filter and one 10.7 MHz FM bandpass filter narrow and shape the upconverted inputs, while a low noise frequency-synthesizer, operating between 11.2 and 11.7 MHz,

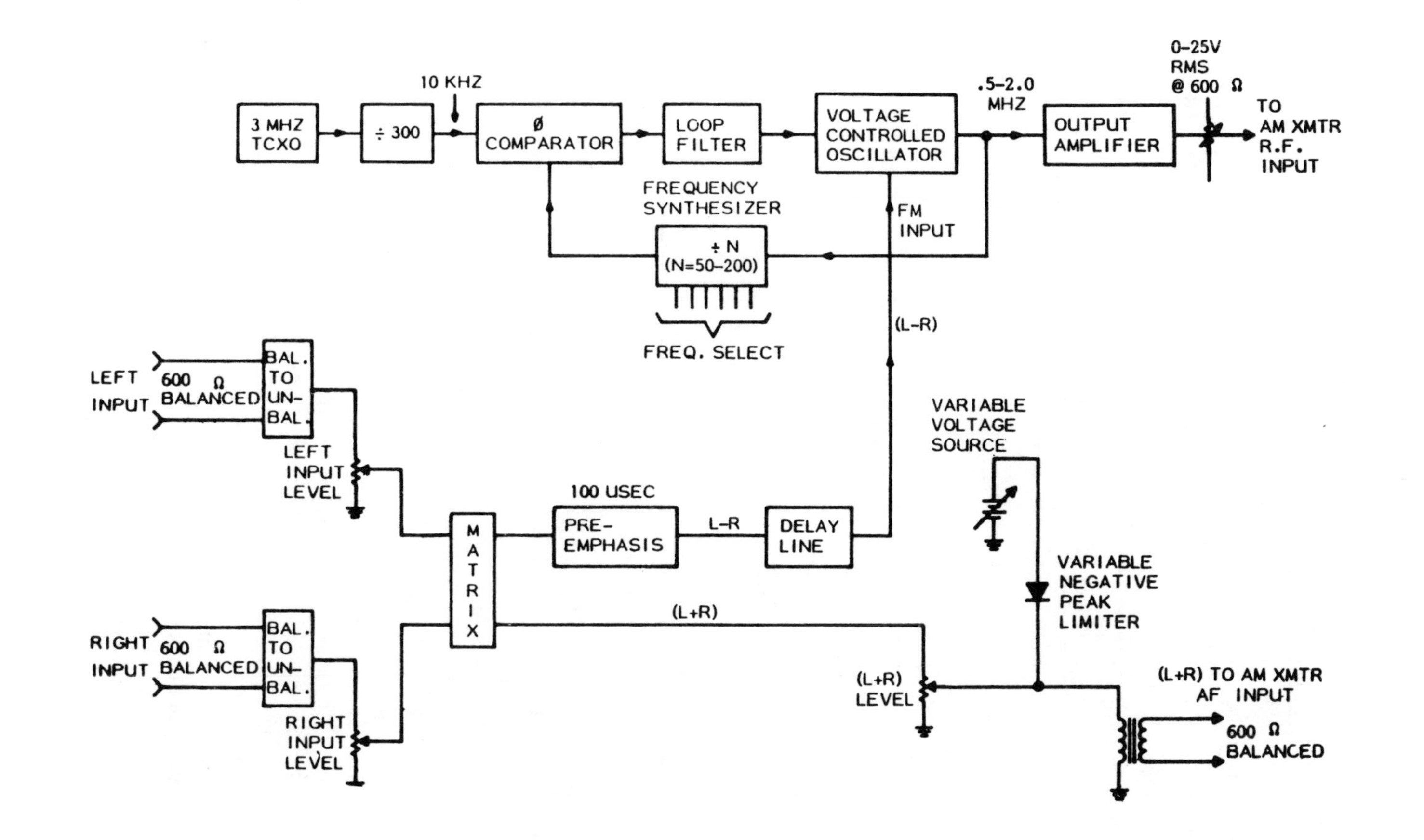

Fig. 1-9. Detailed block diagram of Belar AM-FM stereo exciter (courtesy Belar).

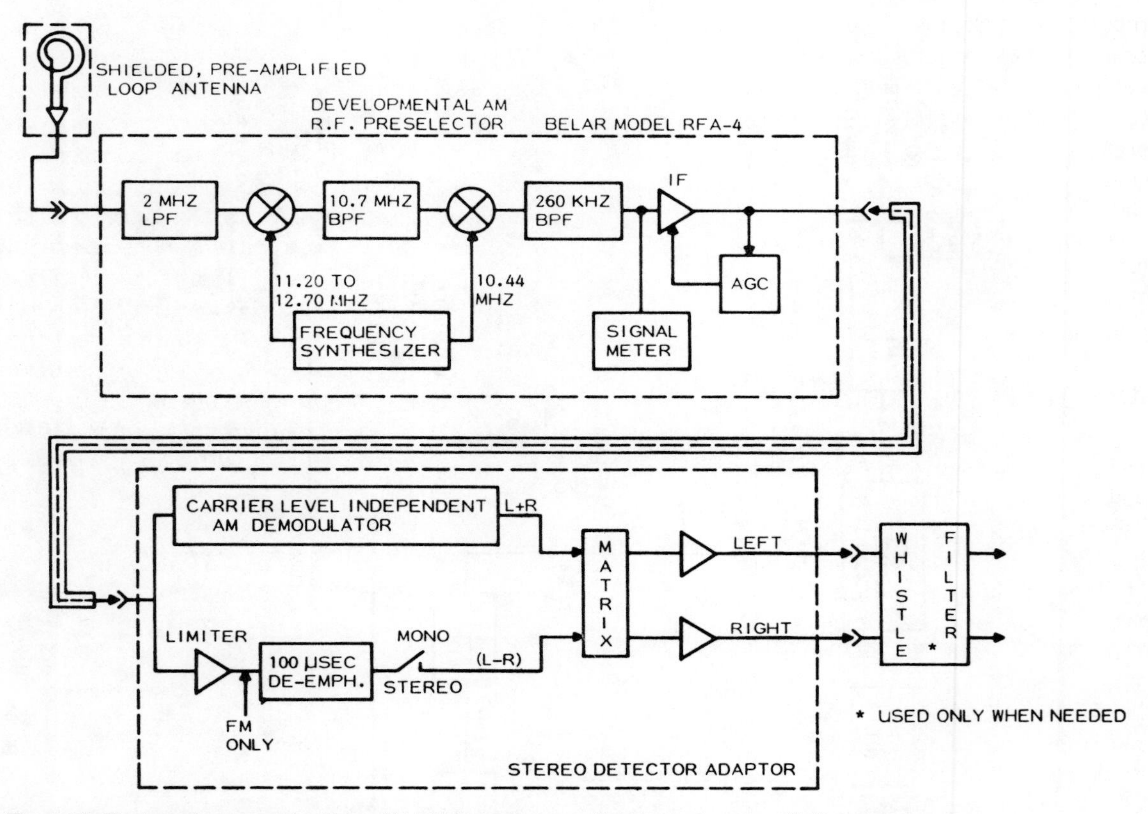

Fig. 1-10. Test receiver used for Belar AM-FM stereo system field tests (courtesy Belar).

represents local oscillator drive of 10.44 MHz, producing a difference frequency of 260 kHz following the second mixer. According to NAMSRC, such upconversion to 10.7 MHz both eliminates image problems and any requirement for "front end preselection." AGC-controlled intermediate frequencies pass into the IF amplifier through a 5-pole inductive-capacitative filter, which determines overall receiver selectivity. IF meter-monitored information goes to the stereo demodulator consisting of an FM limiter and carrier level demodulator, the latter supplying an audio L+R output "which is a function of the ratio of peak envelope voltage to average carrier voltage and is independent of average carrier power changes or carrier shift."

For L–R stereo, the amplitude limiter removes any amplitude modulation, it is then deemphasized by 100 μsec, and quadrature discriminated for introduction to the AM-FM matrix. Here standard left (L) and right (R) signals are separated and then routed through a 10 kHz filter, if required, to remove any adjacent channel interference.

FINDINGS AND RECOMMENDATIONS

The foregoing pretty well concludes the types of transmitters and receivers tested by the various committees of NAMSRC, avoiding some of the excruciating details, but including enough meat to show available AM stereo systems of the late 1970s. In the meantime, of course, there have been considerable developments in both transmitters and receivers, as well as the FCC's declared open market policy of "may the best man win" rather than bearing the burden of selecting a single system. Some of the committee comments in the NAMSRC report, therefore, may be of interest when considering original systems compared with those developed in the '80s. Several of these have now reached commercial applications under actual broadcast conditions. All findings apparently resulted from live radio tests in the Washington, D.C. and North Carolina areas and should, therefore, carry substantial weight in technical evaluations. What follows, then, are selected topics from the AM Stereo Report where fact and utility seem to be coincident.

Stereo Over Existing Transmitters

The Transmission Systems Panel conducted transmit field tests over WGMS and WTOP in the District of Columbia, and WBT, Charlotte, North Carolina. Each had a different model transmitter, but there were no changes to the three stereo exciters except for manual time delay adjustments to match "mid-audio" frequency time delays of the AM/FM/PM channels. Otherwise, an audio jack and crystal oscillator/exciter carrier source were the only other additions. Manufacturers of other transmitters also sent data on their units and it was concluded "there is an excellent chance of successful stereo operation of these transmitters with any of the three stereo systems" (tested).

Actual field tests involved RCA's 10 and 50 kW high level, plate modulated, somewhat antiquated units, and a new 5 kW Harris PDM transmitter. A few "retrofit" circuit development and transmitter modifications did arise with the application of stereo, but *no* problem emerged that would exclude existing transmitters from stereo broadcasting. Among others, here's what the Committee found:

☐ L−R "hum" in one transmitter, probably requiring power supply filtering.

☐ Differences in commons between exciter and the rf chain, necessitating the addition of a coupling transformer and amplifier.

☐ Some stereo distortion and poor separation at higher audio frequencies, probably due to transmitter maintenance, age, and possibly the addition of a modern exciter.

These seemed to constitute the prime difficulties—all of which are easily surmountable—and there were agreeable surprises, too.

Even though existing transmitter carrier amplifiers were not originally designed to handle frequency or phase modulated signals, the FM/PM L−R information in several tests had lower distortion "at all modulating frequencies and lower noise when L+R modulation was not present." And this speaks well for many older designs that are operating in good condition. Further, the Committee, based on its experience, also offered suggestions for future technical refinements in AM stereo transmitters—points which current manufacturers should seriously contemplate.

Future Transmitters

Considering AM stereo may well rival FM stereo in popularity once its use becomes widespread, the Committee suggests built-in stereo capability should be included in future transmitter designs. This, naturally, would or should have add-on stereo generator provisions offering rapid changeover from monophonic to stereophonic at maximum convenience and minimum cost whenever required.

Reasonable bandwidths are necessary too, for anticipated stereo transmissions. Rf stages prior to AM modulation should receive careful attention.

L−R and L+R timings should be matched at *all* audio frequencies "by the addition of all-pass circuits" for envelope delay equalization, or some other equivalent technique that will produce comparable results. Signal generators in closed circuit tests resulted in considerably better L−R and L+R separation than full-fledged field tests with regular transmitters. Therefore, there should be provision for envelope delay versus modulating frequency equalization.

There is also a need for linear-phase carrier-frequency amplification in new transmitters equipped for AM stereo as well as special neutralization of AM modulated or linear amplifier stages. This is needed to minimize incidental phase modulation resulting from amplitude modulation, with loss of stereo separation and subsequent stereo distortion.

Other sources of incidental phase modulation—such as unsymmetrical rf sideband response after AM modulation, antenna bandwidth and loading and nonlinearities of high power tubes—should be taken into consideration.

The Committee's final recommendation suggested that the FCC should consider "either a loose standard initially or an interim standard that will permit stereo broadcasting with less stringent requirements for frequency response, stereo separation, noise and distortion than those imposed on FM broadcasting."

As you will see in the next chapter, the FCC did rule for more liberal AM regulations, but also established minimun technical rules preventing AM stereo interference to other spectrum users as well as requiring acceptable stereo performance. Covered also were mono receiver compatibility, compliance with international agreements, emissions limitations, stereo separation, and adjacent channel protection ratios. So "a free marketplace" did not mean that everyone could get on the air with whatever contraption was handy. There are rules to abide by as well as transmitter type acceptances which must be satisfied before legal airwaves entry begins.

AGC

Forever known as automatic gain control (AGC), the problem in stereo is how to control L+R as well as L−R and still not affect either by unnecessary limiting or outright clipping. This means a well-designed circuit that acts on both AM and FM/PM simultaneously, taking the signal variations of each into account. You could have independent systems, but that wouldn't do; either AGC actions might control both AM and stereo, but this wouldn't help peak limiting. Summing the two signals for auto level control might work out, and certainly separate controls could be established for both sum and difference signals, but this has problems because of stereo gain and phase changes. The final compromise, however, may be a tethering of the two control circuits, but only unidirectionally so that high level sum signals cause dual channel gain reduction to prevent image modulation, with difference signals not affected. The Committee suggests some image modulation may result from this method, but it's a good compromise.

Auto Polarity

The Committee recommends that any device that automatically switches audio signal polarities has to operate on both the sum and difference channels at the same time when the signal is low. There are automatic devices now that use the highest peak level for positive modulation because many program signals are asymmetrical. Ordinary limiters usually clamp high peak signals, resulting in an undesirable loudness imbalance during programs. Such a device should offer maximum signal handling without undesirable clamping.

Phase Shifting

This has to do with all-pass networks reducing program peaks where a circuit increases phase shift with frequency. When all signal harmonics add for maximum peak levels, audio-channel phase changes can modify this harmonic-time function resulting in lower peak signals. By so doing, broadcasters may increase average station modulation—which is, according to the Committee, "good." But, they continue, such all-pass networks should operate prior to final peak limiting and where close-tolerance phase tracking is maintained between channels.

Telco Standards

These allow program line levels up to + 8 VU, and the Committee encourages this practice to maximize S/N ratios since equalized line attenuations may be as great as 40 dB. With an 8 dBm signal, the following *minimum* line standards are encouraged:

Line Grade	S/N
5 kHz	62 dB
8 kHz	62 dB
15 kHz	65 dB

30 dB Separation

The Committee endorses the FCC Rules for a

minimum stereo left-right separation of "approximately" 30 dB. Although this is difficult when working with disc and tape sources, NAMSRC feels it should be maintained.

Studio-Transmitter Links (STL)

In FM, separate left and right audio transmissions may occur on the same rf channel, but with a 100 kHz offset between the two, for delivery of a composite signal. In AM, left-and-right or sum-and-difference audio may occur between the two; but, regardless, the Committee still recommends frequency offset for best separation and minimum crosstalk. There may also be a possibility of reducing the 500 kHz/channel authorized bandwidth which is not required for either AM or FM stereo operation. Nonetheless, the STL signal should operate with 1 percent or less distortion, S/N ratio of better than 55 dB, 30 dB separation, and frequency response between 30 Hz and 14 kHz.

Considering all available STL transmission methods, Panel 2 of the NAMSCR group recommended in the 1977 Report that two separate radio links seemed to be the most "attractive." Both channels, members said, are treated identically, no special multiplexing equipment is needed, while S/N and channel separation are "maximized." In an FM system, narrowing bandwidth requires lessening the audio frequency modulating rate. And when deviation is reduced, the S/N ratio suffers and bandwidth does not contract proportionally.

True, a separate STL link can offer wideband audio, small distortion, and 55 dB S/N in an occupied bandwidth of 50 to 100 kHz. But some STL receivers will be located with an STL transmitter, so receiver overload or channel adjacency problems may be expected in some instances if narrowband channels are allocated for AM stereo service. So present 250 kHz FM channels for one of the dual FM links could accommodate dual AM stereo withour difficulty. However, the allocated 947—952 MHz spectrum for STL service may be inadequate for both FM and AM STL requirements and consideration should be given toward enlarging the spectrum.

Receiver Compatibility

The Receiver Panel picked five types of monophonic receivers for compatibility testing which included: inexpensive pocket portables, multiband portables, some in consoles with record chargers and/or tape players, component hi-fi receivers or tuners, and auto radios. The inexpensive sets have minimum sensitivity and audio fidelity; multibands have more amplifying stages and tuned circuits; consoles reproduce relatively wide audio ranges because of cabinet baffling; hi-fi sets have good tuners and wide system bandwidths optimized for best performance in metropolitan areas with strong signals; the auto radios must handle great varieties of signal strengths and temperature ranges. Here, wide bandwidths often must be compromised for better adjacent channel selectivity and reduction of power line and ignition noises.

Of all the statistics compiled, which include sensitivity, distortion, frequency response, AGC figure of merit, distortion versus modulation, etc., the mean 6 dB (half voltage) bandwidth appears the most interesting. These are taken from a report of A.L. Kelsch, Vice Chairman, Panel III, Receiving Systems, NAMSRC. Pocket portables measured 6.8 kHz (4-11 kHz); modular consoles, 8.5 kHz (5.1-11 kHz); component, 7.2 kHz (5-10.5 kHz); and automotive, 7.3 kHz (6-8.7 kHz). The multiband portables were not covered. At 20 dB, of course, bandwidth responses were virtually doubled, as you might expect.

Stereo Receiver Considerations

AM stereo receivers, as their monophonic counterparts, will have audio responses as a function of bandwidths. But wide bandwidths open the door to additional adjacent channel interference and noise, especially from weak stations. Therefore, expect most stereo receivers to have compromised selectivity (like equivalent AM sets), and certainly crystal-controlled electronic tuners, since precise tuning is required to maintain selectivity and reduce distortion. Fast sampling time for stereo identification could also be desirable, while PLL lockup could be improved so that it is completed more quickly than now.

Table 1-1. Name RC Field Tests.

Monitor system performance . . .	Transmitter stereo encoder audio test and rf generator into wide-band ideal detector.
Receiver system performance . . .	Transmitter stereo audio tests encoder and rf generator into receiver.
Monophonic compatibility . . .	Transmitter stereo encoder and rf generator audio measurments into compatibility receivers.
Envelope modulation limits . . .	Stereo transmission amplitude modulation limits.
Stereo receiver noise . . .	Sensitivity at L+R and L,R, only at 18, 24, 30 dB and S+N+D/N+D.
Occupied bandwidth . . .	Transmitter stereo encoder into rf generator spectrum analysis modulated with four tones: AM L+R, L−R, L,R.
Protection ratio . . .	Noise compatibility receiver measurements from stereo and mono on 1,2,3 adjacent channels.

The big battle in stereo receivers, of course, takes place over methods of detecting stereo in the various systems. In the meantime, the various companies will undoubtedly modify some of the existing radios such as Pioneer's SX6—to accommodate L and R reception and detection. Of course, Delco, Ford, and Chrysler have their own facilities and will continue with whatever radios they choose to place in their automobiles. Specialized designs may take a little longer until the national dust settles or until someone comes up with a device to do more with less and cost virtually nothing.

Field Tests

NAMSRC also developed a series of field tests for which some statistics are not available in the December 1977 report, and therefore they will be listed but not discussed. See Table 1-1.

Groundwave and skywave tests were also conducted over WGMS and WTOP, and WBT, respectively.

System Equations

Magnavox has also offered full system equations for the five *original* systems as proposed to the Federal Communications. These we will happily print without comment (Figs. 1-11A and 1-11B) since you already know the source, and that source has done a great deal of work in aiding the AM stereo cause and also suffered a severe disappointment in so doing. We hope their efforts will eventually be rewarded in many ways.

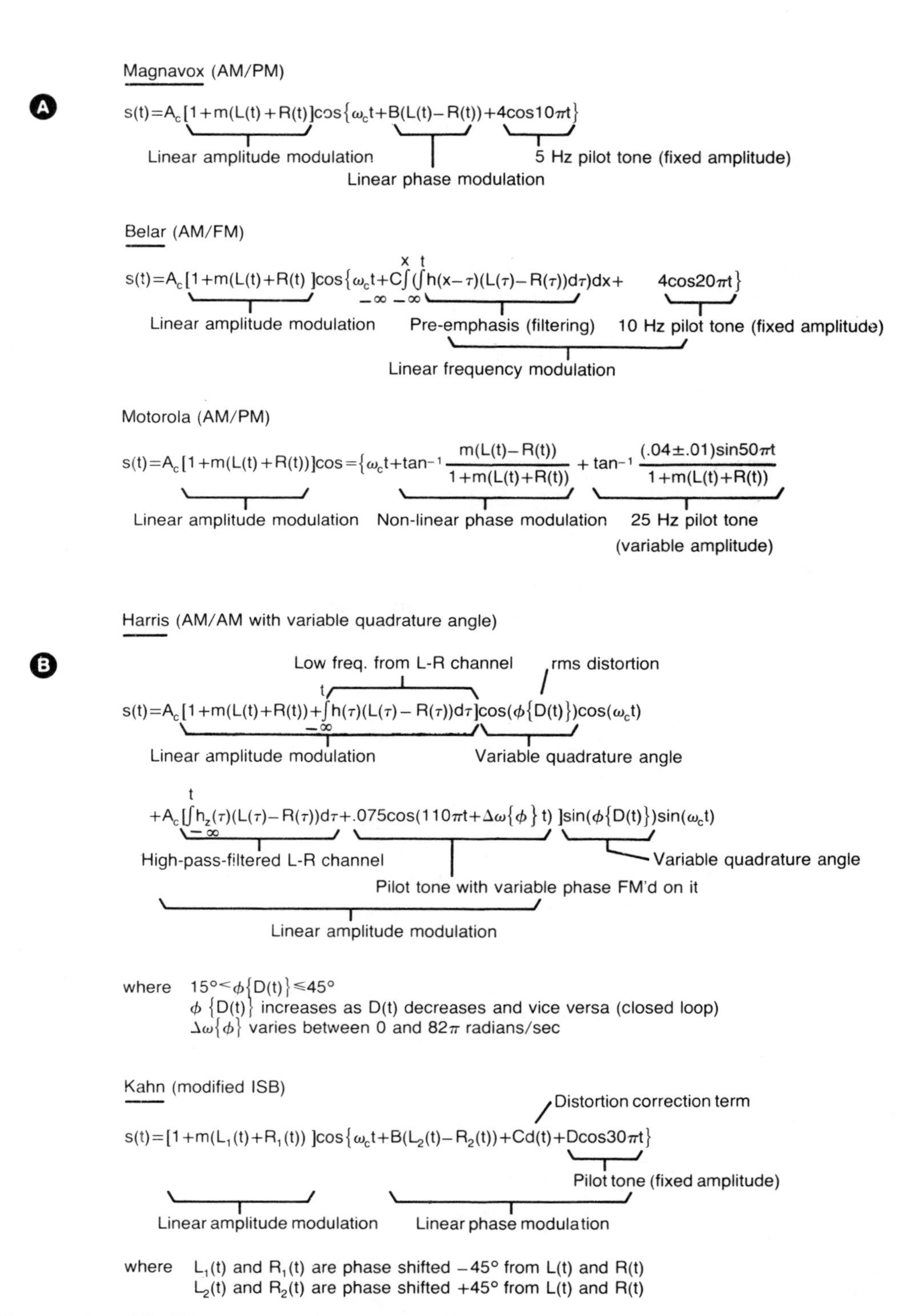

Fig. 1-11. A and B. AM system equations (courtesy Magnavox).

Chapter 2

The FCC's Marketplace Decision

The Federal Communications Commission adopted a Notice of Inquiry on June 22, 1977 following receipt of two petitions for AM stereophonic broadcasting, one proposing a specific system. Following this notice, four additional petitions were received from other sources who had developed their own systems. All five could be adapted to existing AM broadcast transmitters, but each required dedicated receivers for adequate reception. Based on considerable interest by station operators, consumer electronics people, and broadcast equipment manufacturers, but little public response, a Notice of Proposed Rule Making was issued on September 14, 1978.

CONSIDERATIONS

Because none of the systems was obviously superior to the others based on available information, plus certain reservations relative to operation and performance, the Commission sought additional data before proceeding further. A group sponsored by the Institute of Electrical and Electronics Engineers, —Electronics Industries Association, National Association of Broadcasters, and the National Radio Broadcasters Association—called NAMSRC, was formed to conduct comparative studies and tests for the FCC. The three systems tested were products of Belar Electronics Corp., Magnavox Corp., and Motorola Corp. Broadcast equipment manufacturer Harris Corp. participated in the tests but did not submit its own equipment at the time. Kahn Communications, one of the initial developers of AM stereo, did not work with the rest, although, like Harris, offered its system for consideration to the Commission. Kahn and Hazeltine later joined together and their system became known as Kahn/Hazeltine.

Important considerations for this AM stereo system included:

- ☐ Compatibility with existing AM broadcast receivers.
- ☐ Compliance with existing AM broadcast bandwidth limitations.
- ☐ Compatibility with existing AM transmitters and antennas.

☐ No service area loss or loudness for either monaural or stereo reception.

☐ Simple design and reasonable receiver cost.

☐ Satisfactory stereo service for night time skywave reception.

☐ Simple administrative procedures for beginning stereo upon Commission approval.

That such testing and FCC consideration proved helpful was demonstrated by prompt changes in the equipment of three manufacturers.

Belar adjusted its modulation index to reduce bandwidth upon high modulation single channel conditions, Kahn/Hazeltine and Belar added pilot signals for rapid mono/stereo switching, and Harris included a variable frequency companding control pilot for better stereo S/N ratio and improved stereo area coverage. The FCC, itself, soon discovered that occupied bandwidth depends on frequency response to the stereo signal, uniform AM signal loudness would not necessarily allow full stereo separation during certain programs, and mono compatibility often compromises stereo separation. In addition, after listing proposed operating parameters for each of the five systems, the FCC complained that responses should have contained enough information to make a final system selection, but did not. It noted that Kahn/Hazeltine said its system could offer AM stereo immediately, Motorola featured low distortion, and Harris was concerned with transmitted signal bandwidth by audio processing stations during stereo programming. The Commission did note that Magnavox and Belar used the simplest transmit and receive technology, and the Magnavox pilot signal could be used for either data or control transmissions.

Commission staffers urged the Commission to select a single AM stereo system based on available information which was still incomplete. They suggested that a "marketplace" general systems approach would delay system introduction because broadcasters and receiver manufacturers would be reluctant to make a substantial investment in a technology that might not ultimately be successful. Consequently, on April 9, 1980, The Commission decided that the selection of a single system would best serve the public interest and chose the Magnavox system on the basis of an AM stereo system evaluation table prepared by its engineering staff.

However, when the Commission staff was directed to prepare a Report and Order to carry out this decision, a flood of objections was received from broadcast licensees. They didn't like a possible loss in mono loudness when transmitting stereo as well as "popping" noises produced by the stereo receiver during negative peak modulation. They also didn't like the idea that the Magnavox system would require a reduction in audio high frequency fidelity for both mono and stereo reception—a comment that was untrue and actually related to auto ratio nighttime problems of co-channel interference that might be aided by limited bandwidths but could be helped with other methods, too.

EVALUATIONS

The end result produced a proposed quantitative assessment based on analytical numbers rather than qualitative judgments. But the Commission found that testing methods and data documentation varied among respondents, making direct comparisons across sytems difficult or impossible. Therefore, additional information was required before choice of any single system could be attempted. Consequently, a Memorandum Opinion and Order and Further Notice of Proposed Rule Making was prepared and issued on September 11, 1980. This notice contained a revised but incomplete Evaluation Table and explanation of methods chosen, with a request to comment on certain categories and scoring methods used and critiques of the Evaluation Table itself. Table 2-1 is presented as received from the FCC. You may judge its contents for yourself.

Appendix A contains a summary of the criteria used by the FCC in this scoring. A good deal of it may become useful as a guide for future dealings with the Commission or for historical reference when reviewing the anticipated expansion of this particular industry. Once you experience AM stereo, monophonic can't compare.

Table 2-1. FCC AM Stereo Evaluation.

	EVALUATION CATEGORY (INITIAL)	MAGNAVOX	MOTOROLA	HARRIS	BELAR	KAHN
	Numbers in parentheses ( ) indicate the maximum possible scores in the various categories or sub-categories					
I	MONOPHONIC COMPATIBILITY (15)	**12**	11	7	12	11
II	INTERFERENCE CHARACTERISTICS					
	(1) Occupied bandwidth (10)	**7**	5	9	5	8
	(2) Protection ratios (10)	**5**	3	8	5	7
III	COVERAGE (10)	**7**	6	6	5	5
IV	TRANSMITTER STEREO PERFORMANCE					
	(1) Distortion (10)	**8**	7	3	9	2
	(2) Frequency response (10)	**9**	4	5	10	7
	(3) Separation (10)	**9**	9	6	10	2
	(4) Noise (10)	**7**	8	7	6	6
V	RECEIVER STEREO PERFORMANCE					
	Propagation degradation (5)	**3**	5	4	3	5
	Directional antenna effects (5)	**3**	3	4	3	3
VI	MISTUNING EFFECTS (5)	**3**	3	4	3	3
	TENTATIVE TOTAL SCORES (100)	**73**	64	63	71	59

SEPTEMBER 9, 1980

Linear and Nonlinear Systems

Staff engineers also wanted to evaluate at least part of the overall system design and chose to discuss trade-offs between linear and nonlinear systems. Parts of this are especially interesting considering various differences and trends among AM proponents and manufacturers.

Two equations, mathematically equivalent, are given for carrier modulated signals:

$$y = A\,\mathrm{Cos}(W_c + \phi\text{- or } y$$
$$= M(I)\mathrm{Cos}(W_c t) + M(Q)\mathrm{Sin}(W_c t)$$

Where: Wc = carrier frequency

A = signal amplitude

Q = signal phase

The following definitions apply to either simple or complex functions.

M(I) = coefficient of in-phase component (of the modulating frequencies)

M(Q) = coefficient of quadrature component (of the modulating frequencies)

One or the other may be used interchangeably, but Staff recommends the one more appropriate to "underlying physical phenomena." The second equation is most applicable to sidebands generated by the modulated carriers. Should M(I) and M(Q) consist of either simple or linear sums of simple tones, then the system is linear.

	FINAL EVALUATION CATEGORY	MAGNAVOX	MOTOROLA	HARRIS	BELAR	KAHN
	Numbers in parentheses ( ) indicate the maximum possible scores in the various categories or sub-categories.					
I	MONOPHONIC COMPATIBILITY					
	(1) Average Harmonic Distortion (15)	**15**	9	6	9	12
	(2) Mistuning Effects (5)	**5**	5	5	5	5
II	INTERFERENCE CHARACTERSTICS					
	(1) Occupied bandwidth (10)	**3**	4	10	5	6
	(2) Protection ratios (10)	**7**	7	8	1	9
III	COVERAGE (Relative to Mono)					
	(1) Stereo to mono receiver (5)	**5**	5	5	5	5
	(2) Stereo to stereo receiver (5)	*	*	*	*	*
IV	TRANSMITTER STEREO PERFORMANCE					
	(1) Distortion (10)	**8**	8	6	8	4
	(2) Frequency response (10)	**8**	5	5	6	8
	(3) Separation (10)	**10**	10	10	8	3
	(4) Noise (10)	**6**	10	8	6	8
V	RECEIVER STEREO PERFORMANCE Degradation in stereo performance over that measured at the transmitter, including consideration of directional antenna and propagation degradation (10)	**9**	8	9	9	5
	TOTAL SCORES **MARCH 18, 1982**	**76**	71	72	58	65

$$M(I) = A + \sum B_m \text{Cos}(W_m t),\ M(Q) = C + \sum D_n \text{Cos}(W_n t)$$

Where W_m and W_n are simple tones and A, B_m, C and D_n are constants and independent of frequency.

This shows, according to Staff, that with no harmonics or intermodulation products, a linear system can be generated, and all simple tones will lay within a single set of sidebands on either side of the carrier. Naturally there will be no distortions or out-of-band emissions and only simple sum and difference frequencies can appear in the sidebands.

But when modulations are nonlinear, both intermodulation and sideband products of higher harmonics result, and to offer the same frequency responses as a linear system without distortion, nonlinear systems must be at least double in bandwidth. With the 30 kHz bandwidth of AM systems, the maximum audio frequency at which stereo separation can be attained is, at best, only half that of a linear AM stereo system. The FCC Staff used these criteria to help evaluate the various transmit-receive system involved as follows:

Harris. Its transmitter frequency response is 50-15,000 Hz in the L+R mono channel and 200-15,000 Hz in the L−R stereo channel. Below 200 Hz, Harris uses the L−R channel to transmit a compander control tone for stereo improvement while "maintaining compatibility with envelope detector receivers." However, should envelope detectors be replaced by synchronous detectors (as is now becoming apparent), Harris can revert to a fixed rather than a variable pilot tone. Then an L−R range of some 100 Hz to 15 kHz could be made available to transmit stereophonic intelligence.

Magnavox, Motorola and Belar. These systems, according to Staff, are limited to 50-7,500 Hz in the L−R channel, and Kahn/Hazeltine to but 50-5,000 Hz. These are inherent features of the four systems and, according to the FCC, "cannot be improved within the existing bandwidth restriction." And in the future, if international agreements force reduction of the maximum. AM modulation frequency below 15 kHz, L−R response would have to be reduced proportionally, leaving Harris still having the best stereo channel moduation from its transmitter.

In receiver design, Harris shows major strength without distortion in the recovered signal in that its decoding process is precisely the inverse of the transmitted intelligence. The receiver then, would theoretically not introduce distortion in decoding. In the other four, according to Staff, "hard limiting in the receiver causes inherent distortion in the decoding process." On the other hand, Harris requires a more complex pilot tone detector since the encoded L+R and L−R ratio is deliberately altered before transmission and the pilot tone detector must determine the level of such change and deliver a correction voltage to the matrix decoder. Harris counters, however, that even though some separation is lost, there is little detectable change in its signal when receiving at fixed gain reduction of 1/1.414 in the L−R channel.

As the FCC points out, detected L+R and L−R signals are matrixed in AM stereo to produce L and R information. To prevent crosstalk, it is essential that the ratio of the L+R and L−R signals at the input to the matrix decoder be the same as their ratio at the output of the transmitter decoder. If there are either rf or IF gain changes affecting one channel and not the other, the decoder inputs must compensate accordingly. Staff observes that Harris uses two identical detectors for both channels, but Magnavox, Motorola, Belar and Kahn/Hazeltine each use an envelope detector in the L+R channel and some form of angle detector in the L−R channel. Therefore, gain compensation circuits are needed to restore the correct L+R and L−R ratio at the input to the matrix decoder.

Staff once more emphasizes that intermodulation products are apparent in all nonlinear systems, and the high resolution NAMSRC 4-tone test showed 400 Hz intermodulation for Magnavox, Motorola, Belar and probably Kahn/Hazeltine had that system been tested. Further, Motorola and Kahn/Hazeltine have only moderate compatibility with synchronous detectors, while Magnavox and Belar have practically none at all. Harris, of course, is the most compatible. Sansui, a developer of AM synchronous detection, believes, eventually, all AM receivers will utilize this technology and that the envelope detector will disappear just as the vacuum tube receiver." Sansui does note that linear quadrature systems produce some distortion in mono receivers using envelope detectors, but such distortion is *less* objectionable than from bandwidth limiting.

Of course, a typical diode detector will introduce negative modulation distortion peaks because of its obvious nonlinear characteristics. Staff, therefore, says that Magnavox, Motorola, Belar, and Kahn/Hazeltine, cannot achieve 100 percent negative modulation and still maintain stereo separation during the negative peak modulation interval. When amplitudes go to zero, phase information is necessarily lost.

Obviously, if synchronous detectors replace envelope detectors in AM stereo and mono detectors can function well, a variable pilot tone will no longer be necessary, and a linear system such as Harris can revert to full quadrature. "The pilot tone would be fixed on its quiescent frequency of 55 Hz and the frequency response of the L−R channel could be extended down to 100 Hz," according to analyzing engineers.

Further Notice Replies and Decision

Thirty-three rejoinders were received to the September 11, 1980 Further Notice of Proposed Rulemaking, 23 of which were formal comments and 17 replies. Belar did not respond. The Commission noted that no "new matters or issues of significance were raised." But there were criticisms of measurement ways and means, and a large number of broadcast licensees informally voiced concern that "the selected AM stereo system must not require any reduction in modulation loudness, high frequency response, or area coverage." Many comments urged the Commission to allow the marketplace to settle the selection of the AM stereo system. Hazeltine used video recorders as an example and cited the many different and incompatible technologies that were originally offered but that only the "more satisfactory products survived." Hazeltine also believed that the Commission had a greater probability of being incorrect than the marketplace "and that, if more than one AM stereo system survived, manufacturers of consumer products would readily meet the demand with receivers capable of decoding more than one stereo system."

This, of course, has already happened to a certain extent with Sansui, and even at this writing, others have integrated circuits, either now or in development, to decode any of the four extant systems with varying results. So certainly in this respect, outside advice was well heeded. Both the National Broadcasting Company and the American Broadcasting Company concurred in a free marketplace, but Sony and the Electronics Industries Association weren't so positive about 4-way decoder chips. Both thought that receiver production costs would increase substantially.

The FCC then considered a lottery to determine a system, but this was discarded because respondents weren't in favor. So after assigning two senior "technical personnel" (engineers) to the project full time for five months during the review, the Commission found that "any decision for one AM stereo system would be highly tenuous." In addition, "data possessed by the Commission are incompatible in some instances since no uniform test procedures were employed. Second, the weights assigned to the various factors and the engineering judgements employed are subject to variance, depending on the analyst. Finally, the results obtained are close even if the data and methodological difficulties were absent. Thus from the results in the evaluation table, no clear choice is apparent in any case."

And so the Commission made its final "free marketplace" ruling on March 4, 1982, with only Commissioner Abbott Washburn dissenting. Between then and now, of course, the market shakedown has taken place. The automobile manufacturers, too, are beginning to advertise mobile AM stereo along with FM stereo, and progress should now become much more rapid in developing this long overdue service. In the meantime, we'll continue to gather the latest information for print right up to the time this book appears on the market.

You must also realize that all the foregoing has to be considered preliminary information leading to market and technically induced changes that have taken place since initial FCC tests and filings. Some systems have even been selectively changed and present somewhat different electronic execution and format than originally intended. All of this will be explained at length in succeeding chapters since AM stereo manufacturers would like maximum exposure for their respective systems.

FCC Regulation Changes and Additions

You may find the FCC regulation changes of either some immediate use or for extended reference. Therefore, they have been included in Appendix B. While not monumental, they are specifically oriented toward AM stereo, and broadcast equipments will have to comply. There are also some definitions of significance scattered throughout those pages that can be worthwhile. There's no editing involved since various regulation sections are referenced and the FCC wants its own language in Federal requirements and instructions. To the best of our knowledge, all FCC actions published here are still in force.

Chapter 3

Kahn/Hazeltine and Harris Systems

Harris and Kahn/Hazeltine AM stereo systems are placed in a chapter by themselves since their equipment was not tested by the National AM Stereophonic Radio Committee (NAMSRC) prior to its report of 1977. Information about these early systems and theories of operation, however, has become available from the manufacturers and other sources so that a reasonable description of each can be presented and should supply substantial information. We'll try and refrain from final equipment updates until Chapters 4 and 5 so that you will have any comparisons between the old and new involving the four active systems as they operate, or hope to operate, today. At least you'll receive all the information that's not proprietary up to the time the manuscript reaches the publisher. We'll begin with Harris and conclude with Kahn/Hazeltine since some further information on the latter may become available later. See the end of the chapter for more information on Harris.

HARRIS AM STEREO

In its May 15, 1979 comments on proposed rulemaking to the Federal Communications Commission, Harris claimed that it satisfied the following *essential* criteria for AM stereo:

- ☐ Stereo transmissions must be compatible with the existing allocation structure to prevent any increase in AM interference.
- ☐ There must be a compatible signal to existing AM receivers.
- ☐ Inexpensive receivers should have acceptable performance.
- ☐ AM sets should be priced to reach the mass market.
- ☐ Stereo transmissions must be compatible with foreseen technical changes in AM broadcast service.

Of the five proposed systems, the comments said, Harris is the only system which satisfies all these criteria. Harris continued that since its system is a linear AM/AM transmission with bandwidth and sidebands identical to monophonic radios, it would generate neither adjacent nor co-channel interference, and would fit future developments as well as existing monophonic ones.

On March 14, 1979 Harris revised its 30-degree fixed angle between left and right channel sideband vectors to vary this angle so that S/N ratios are optimized while still maintaining the same degree of compatibility with monophonic receivers. Harris calls this the Variable-Compatible Phase Multiplex system, which is says "dramatically" increases the system's stereo geographical-spread. According to Harris, it now exceeds 90 percent of the mono coverage area, and has "excellent" channel separation. Later, a 90-degree fixed angle was adopted.

Theory of Operation

This system is known as an AM/AM linear or additive type since transmissions are the *sum* of *both* amplitude modulated signals, and are transmitted on the same frequency. Competing AM/PM or AM/FM systems have their information *multiplied* together for the complete stereo signal, and are said to be "nonlinear," according to Harris.

Left and right stereo information commences on two separate carriers, in modified quadrature, at the same frequency, but separated by a phase angle that may vary between 30 and 90 degrees. This is done to maximize signal-to-noise ratios as program content changes, "while at the same time maintaining a high degree of compatibility" with monophonic envelope detectors. An FM modulated, low frequency pilot tone tells the stereo receiver how and when to track the instantaneous phase angle so it can follow broadcast transmissions.

These begin with two carriers of the same frequency out of phase by 90 degrees. Both are separately amplitude modulated with left and right audio signals as illustrated in Fig. 3-1. They are then added together, producing a complete VCPM envelope shown in Fig. 3-2.

Whenever audio program content would result in envelope detector receiver distortion, the signal is made to deviate from full quadrature and the phase angle between the two carriers is automatically reduced to a degree that's compatible, maintaining envelope detection in receivers at an acceptable distortion level. Phase angles are reduced only when there is high modulation and source material requires considerable channel separation.

Then, any extra noise generated by departure from full quadrature is "masked," according to Harris, by loud signals "covering up" the noise. This

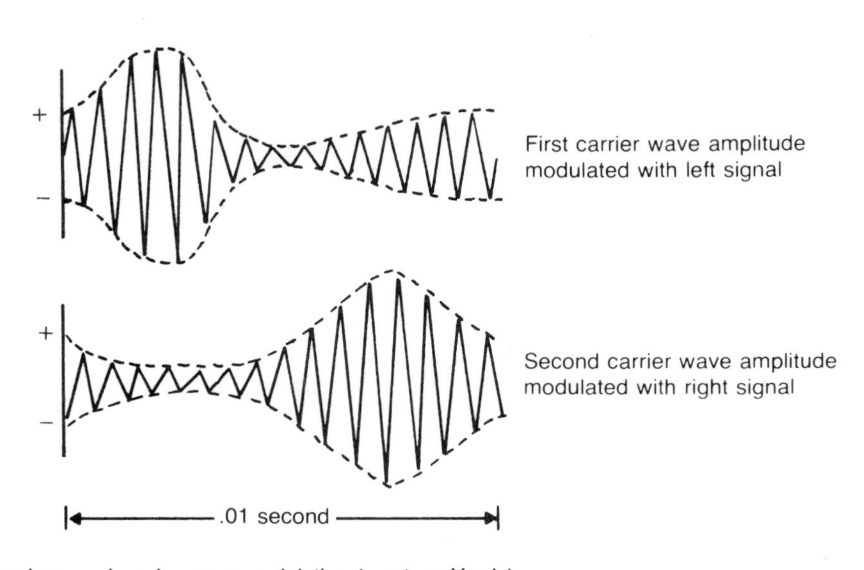

Fig. 3-1. First and second carrier wave modulation (courtesy Harris).

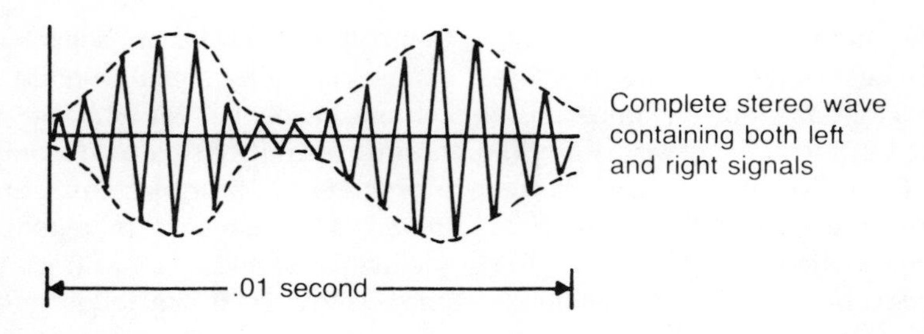

Fig. 3-2. Harris' complete VCMP signal (courtesy Harris).

phase angle between the two carriers is constantly changing, says Harris, optimizing S/N in stereo receivers and their envelope detectors. Such phase angle tracking is delivered to the receivers by a low frequency pilot tone that varies between 55 and 96 Hz, and is proportional to the changing angle between the left and right carriers. Pilot tones also switch receivers automatically to the stereo mode and light a stereo indicator lamp.

The time required for a Harris receiver to identify stereo, light the lamp, and swing to stereo mode is known as "lock time," which is allegedly a rapid 50 milliseconds; and the fastest, according to Harris, of any of the competing systems. (According to industry sources, this lock time has now been lengthened.) Obviously, this would be especially applicable to automobile radios where slow or inadequate tuning could prove distracting to any vehicle operator.

The pilot tone, meanwhile, is protected from interference by "eliminating channel separation" in audio less than 200 Hz, "while still retaining the mono portion of the signal. The ear does not detect this filtered loss since very low frequency sound components don't contribute to spaciousness or the effect of stereo separation.

VCPM Receivers

Variable-Compatible Phase Multiplex receivers and others of many types can process the Harris system. Both synchronous and nonsynchronous means may be used. At the moment, a stereo decoder detects the right and left signals, and two amplifiers plus speakers reproduce the sound. Since AM mono and FM stereo combo-receivers already have dual amplifiers and speakers and existing rf tuners are adequate, the only real design change, according to Harris, is replacement of the envelope detector with a stereo decoder.

Such a decoder for VCPM has two detectors to recover the two signals, and may be designed as variable or fixed angle units, depending on whether maximum or minimum performance is specified. In variable angle decoders, the two detectors are synchronized precisely with the modulating carrier wave angle, and channel separation remains at maximum level. The fixed angle set fixes the phase angle at an average of 70 degrees and reduces channel separation somewhat. Harris says, however, that "most listeners will be unable to detect the difference in stereo quality between the fixed and variable angle receivers."

Harris further states that competing nonlinear systems require a bandwidth of some six times the highest audio broadcast frequency, while its system needs only twice the bandwidth and would, therefore, be within the 30 kHz AM station authorization. Sidebands of competing stations, according to Harris, would actually overlap, "making it impossible for the receiver to completely separate the stations." Other systems, says Harris, would have to compromise channel separation, increase harmonic distortion and limit frequency response, "significantly impairing the quality of their broadcasts whether received by stereo or monophonic receivers."

Harris claims that VCPM systems will offer AM stereo service comparable to FM stereo within existing limitations and full frequency range since carriers and sidebands are identical to mono AM transmissions. Harris also alleges that FM and PM systems can only offer outstanding performance when very wide, occupied bandwidths are permitted. The company points out that FM is permitted 240 kHz versus the 30 kHz allowed for AM but, according to Harris, since most program material today is actually high fidelity, AM stations that don't limit frequency response are really transmitting maximum range AM already.

Completing nonlinear systems, Harris continues, require an occupied bandwidth approximately six times the highest audio frequency broadcast, and would need 90 kHz bandwidth to reproduce audio frequencies of 15 kHz. Other systems, of course, do have means to reduce sideband problems as well as adjacent channel interference, and 15 kHz is by no means a constant CW signal. But there may be some channel separation problems above certain frequencies that bear looking into. We should get into that shortly.

But it's already apparent that *all* stereo AM system proposals have some limitations, and whoever corners the market must have both cost and performance compromises that make the package highly attractive. In addition, add-on station and receiver adaptions now may become quickly outmoded when advanced designs are available by or before 1990. In the meantime, AM stereo, like TV stereo, has given transmitter and receiver manufacturers another big boost towards the manufacture of better products for consumers everywhere. Digital radio and TV probably follow, and these should eliminate many noise, secondary imaging, and bandwidth problems that exist temporarily today.

Test Information

Independent test data submitted by the Canadian Department of Communications Technical Advisory Subcommittee on AM Stereo (over CKLW radio), showed that the VCPM Harris system did not alter or distort protection afforded by directional antenna systems and no indications were noted of any significant increases in the received carrier stereo mode or fundamental sideband levels referenced to monaural transmissions. Nor did directional antennas seem to critically affect channel separation, but channel separation below 200 Hz is minimal, as expected, due to the Harris' fourth order double Butterworth high pass filters used to remove the L–R signal components below 200 Hz. Most record companies, the Committee added, mix "little or no separation into their product at these lower frequencies."

Nor did the Harris system produce any added interference to "first" adjacent channel stations and there was no substantial harmonic distortion increases when signals were detected by ordinary mono, band-limited AM receivers. Reasonable stereo signals were also heard where skywave and ground waves were present as well as other areas with skywaves alone where signal strength was sufficient to override interference levels.

Compatibility tests of the system over CKLW indicated that stereo transmissions were as good or better than normal mono and that there was no substantial increase in occupied signal bandwidth between stereo and mono. The Canadians also noted that Harris has now added a low frequency correction circuit, allowing L–R below filter cutoff to be rotated 90 degrees and added to the L+R component so that relative amplitudes of low frequency components are essentially flat and "simply divided between the left and right audio channels." (Once again, industry sources say Harris has as much spectrum spreading as anyone else.)

Signal-to-noise mono-stereo ratios were determined to be within 1.5 dB of one another in this system and did not restrict area coverage or signals for monaural receivers. Synchronous detectors were also said to be advantageous.

KAHN/HAZELTINE SYSTEM

This system, originally developed by Kahn Communications, Inc. of Freeport, New York, is now called the Kahn/Hazeltine stereo AM system as a result of participation by the Hazeltine Corp. of Greenlawn, New York and which, we understand, is now developing a four-system IC decoder for *all* AM stereo receivers.

In this system, called independent sideband (ISB), the left signal is transmitted as the lower frequency sideband, while the right signal contains the upper frequency sideband.

Leonard Kahn believes that when L=R, a conventional monophonic AM "wave" should be produced; that AM stereo should be fully compatible with conventional AM receivers relative to distortion and frequency response; that when L=R and a single transmitted wave equivalent to 100 percent modulation passes through the system, the signal's envelope modulation should also be 100 percent and within 1 percent accuracy. Mr. Kahn points to over nine years of operational experience with his system in both the U.S. and Mexico, especially calling attention to specific *lack* of any adjacent channel interference even among stations separated by only 20 kHz.

He does indicate, however, that when a stereo receiver is used, stereo separation in his system is *not* degraded when the ISB stereo bandwidth is reduced. But when two conventional AM receivers are adjusted to upper and lower sidebands, stereo separation will be degraded if one or both of the receivers are deliberately tuned "closer to the carrier" to avoid adjacent channel interference in weaker signal areas.

In replying to FCC questions, Mr. Kahn saw "no major problem" developing from adjacent channel carrier interference since notch filters in stereo receivers can effectively remove it. Mono listeners should not be able to hear any difference between stereo transmissions from one station and/or interfering stations in any co-channel interference. Inexpensive receivers, he notes, trade fidelity for bandwidth and attenuate adjacent channel carrier beats, causing little difficulty. He also replied that the effective range of his stereo system is the same as monophonic and that there was no loss in monophonic loudness during stereo operation, even in fringe areas. there is a 3 dB loss in signal-to-noise for ISB stereo compared wiht monaural reception. Combined ground and skywave fading should be about the same for normal mono transmissions.

Mr. Kahn said also that he was "unaware of problems with any transmitter that cannot be readily corrected" when adapting AM broadcast transmitters to stereo. Using his system, he claims it should take only two or three hours to install the exciter, but he does note that in early tests, transmitters should be "properly neutralized" to do away with out-of-band radiation and optimize stereo separation. Transmitters should also be reasonably free of undesirable phase modulation. Kahn exciter cost, he estimates is over $6 thousand (and may have increased recently), but this does not cover studio equipment, interconnects, and installation or engineering. In the beginning, Mr. Kahn favors the use of two receivers for stereo reception since these are immediately available and, of course, work with his system.

Theory of Operation

In the beginning, we had some difficulty in interpreting the actual "fine-tooth" operation of the Kahn/Hazeltine system, but enough information has surfaced from various sources to make the description reasonably complete. As usual, we'll start from the beginning with transmitters and work toward eventual signal reception and processing.

Vice President B.D. Loughlin of Hazeltine Corp. calls the Kahn/Hazeltine product an independent (multiplicative) sideband AM stereo system where left information is transmitted by lower carrier sidebands and right information by upper sidebands. L—R modulates carrier phase, while L+R modulates the amplitude of the phase-modulated signal. There are some similarities to phase-separated and AM/PM systems, but there is a "critical" distinction. The 90-degree phase difference between the two audio modulations results in L and R outputs "each being essentially single sideband (SSB). Therefore, transmitted L and R stereo separates into two different frequency bands and are not combined "as in phase-separated systems."

As explained, "the envelope of the transmitted signal is linearly proportional" to L+R, resulting in higher order sidebands which are "predominantly" second harmonics, eliminating harmonic distortion. This multiplicative action is also said to produce a second harmonic phase modulation term "which, if produced to have all the relevant energy on the correct side of the carrier, can be cancelled by introducing a subtracting second harmonic component in the L–R channel. See the transmitter block diagram (Fig. 3-3). As is evident, a negative and positive phase angle between L and R information produces a full 90-degree phase rotation, with feedback from the +45-degree segment into the second harmonic component. Its output is then summed with the L–R. Subsequently, both left and right inputs are combined in the AMP MOD with their necessary frequency separation intact.

When modulation frequencies are sufficiently high, the second harmonic "term" may be attenuated because of receiver IF selectivity, but since the modulation percentage of higher frequencies is usually constrained, any resulting distortion is "acceptably low." With low percentage of inverse modulation, the Independent Sideband (ISB) system "is not limited by extra receiver noise problems" and heavy modulation on one-sided program material doesn't produce adverse effects. While certain phasing conditions must take place in any decoder for adequate stereo separation, separation is *not* affected by any phase variations in transmission. So phase distortion won't affect stereo separation because of separate L and R difference frequency transmissions, but phase distortion can affect distortion cancellation, especially if the second harmonic and fundamental are incorrectly phase related.

It is also true that long multipath delays won't affect stereo separation in the ISB system, but they can affect distortion cancellation. Under worst case conditions, the second harmonic term can be shifted as much as 90 degrees, but amplitude ripples in the propagation path can be as objectionable as distortion cancellation, according to Mr. Loughlin. He notes, too, that random noise in each channel will increase by 3 dB when the L–R channel of an ISB stereo receiver turns on, and a bit more noise will be generated because of distortion correcting inverse modulation. But even at 100 percent downward L+R modulation, "the gain increases by only 6 dB and goes down by 3.5 dB on the upward modulation swing."

Modulated co-channel interference stays on the proper left or right side, since frequency controls the ISB system and not phase. Further, adjacent monkey chatter" will appear on only one chan-

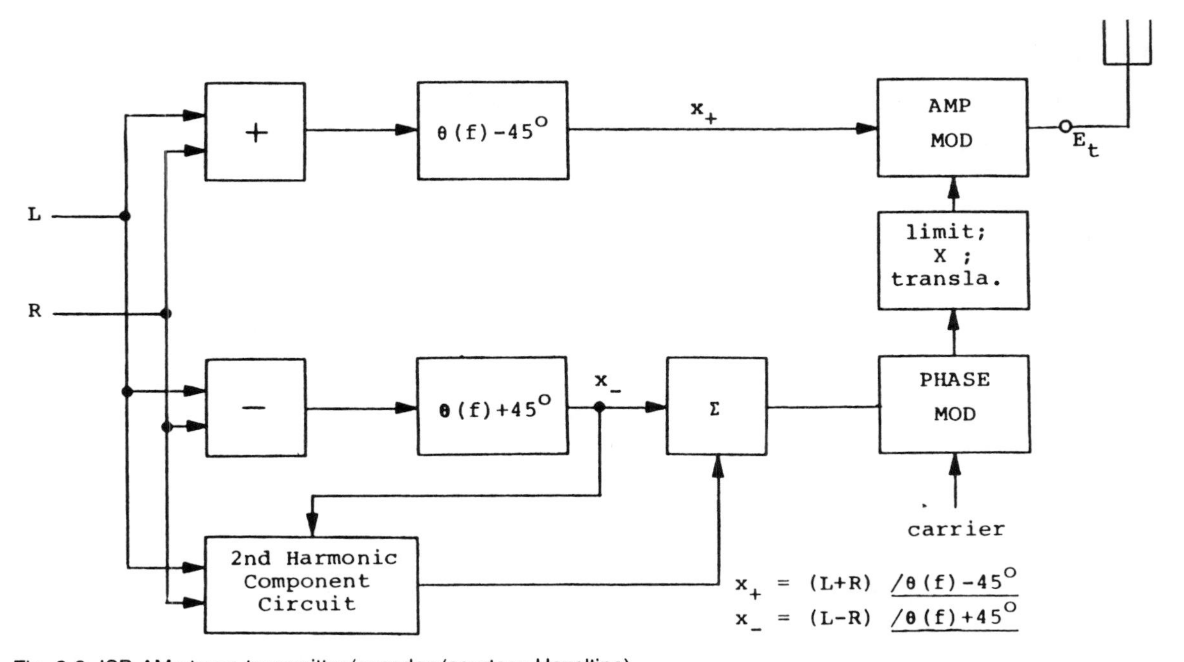

Fig. 3-3. ISB AM stereo transmitter/encoder (courtesy Hazeltine).

nel, depending on whether such interference is low or above the specific channel.

Mr. Loughlin also has some analytical comments on the ISB receiver which we'll paraphase for brevity and simplicity. Perhaps it may help.

ISB stereo receivers may be two regular AM sets detuned slightly "on opposite sides of the carrier," a dual IF receiver, one for L and the other for R, or a receiver that's the inverse of the transmitter with 90-degree phase-different separation. The latter is illustrated in Fig. 3-4.

As you might expect, the L+R information is decoded by an envelope detector, and the L−R signal by a synchronous detector. As usual, there's the phase-locked loop and 90-degree carrier phase shift, along with detection by the "inverse modulator," having appropriate input from the envelope detector. The L−R signal has a distortion component which may be removed by inverse modulation from L+R and requires about 50 percent of the envelope modulation to do it. A quadrature sync detector picks up the 90-degree carrier phase shift and the two signals are then applied to a sum and difference matrix for the final left (L) and right (R) outputs.

SYSTEM CRITIQUES

About the most calculating way to take the starch out of any stuffed shirt is to have his faults examined and discussed by a group of peers. It's also a way to print unfounded and untested criticism, aping more than a few 1980s politicians whose demise will make better news than their emergence. In such practice we do *not* engage, but we would like to excerpt a few pertinent comments from among many, as the hotly contested AM stereo struggle continues without a clear-cut winner (just yet). Which among the four contestants (unless Belar resurfaces) finally triumphs we can, at this stage, hardly say. But Motorola seems to have an edge among the mobile receiver fraternity, while others could gain some ground eventually. Here, only time and the market will tell. But once again, expect the Japanese and other Asians to pick up the lion's share of receiver sales while U.S. types go after the transmitters and encoder/decoder ICs. Regardless, AM stereo is here to stay and flourishing, as we'll report in the fifth chapter. For now, here are a few initial system "rebuttals" as one segment of the AM stereo industry views another.

Kahn/Hazeltine Comments. Mr. Loughlin of Hazeltine says the Motorola system is less sensitive to differential phase distortion than the Harris system, but the latter will likely require very rugged construction in automobile receiver oscillator circuits "to maintain spatial integrity of the stereo

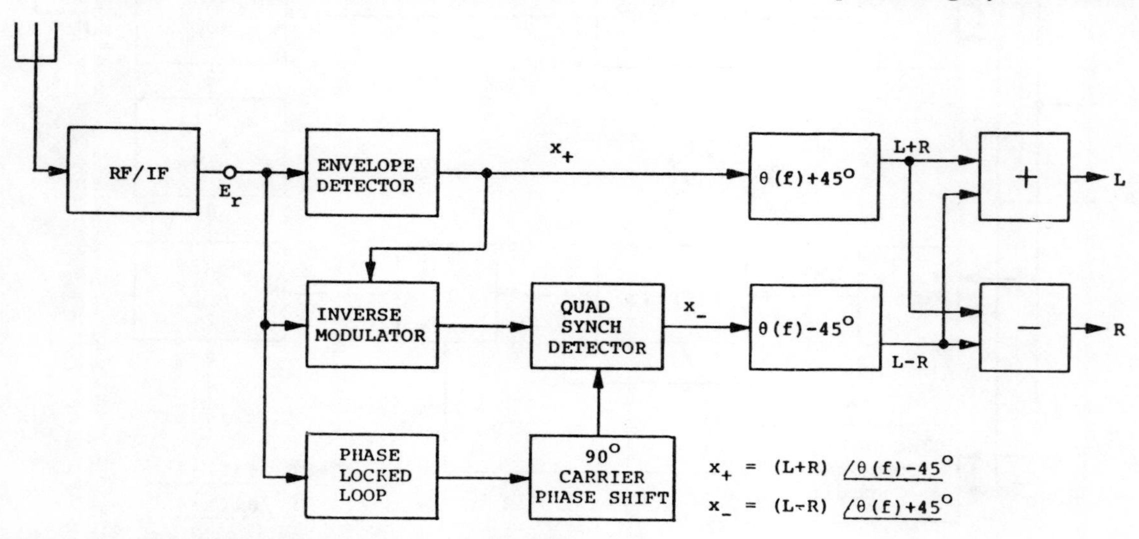

Fig. 3-4. ISB AM stereo receiver/decoder (courtesy Hazeltine.)

program material." He also sees the Harris "off phase" detection resulting in desired signal recovery loss, with degradation in both S/N and "sensitivity of stereo separation to phasing errors." He claims an "extra receiver" noise problem for Motorola and points to "practical limit(s)" on downward envelope modulation due to "extra receiver" noise produced by inverse modulation in the receiver. Mr. Leonard Kahn believes that the Magnavox, Motorola and Belar are unfamiliar with their "click and pop" problems which, he says, are basically signal-to-noise difficulties. These, he said, have been fully described by S.O. Rice of Bell Laboratories "who taught that when the noise component equals the signal component, a frequency modulation detector will produce clicks due to the fact that the vector sum of the signal and noise components swings suddenly from +90 degrees to −90 degrees or vice versa. Assuming the noise and signal are exactly equal, this change in phase theoretically occurs in zero time. Therefore, an impulse function or a click is produced."

Harris' Comments. Nonlinear systems, says Harris, have multiple sidebands and their bandwidths occupy "six times the highest audio frequency broadcast." Kahn/Hazeltine is said to have no first-order upper sideband, but it does have pre-distortion sidebands "appearing asymmetrically on both sides of the carrier." Motorola's predistortion sidebands decrease monotonically with increasing order at a higher rate, but its higher-ordered sidebands decrease more slowly. Therefore adjacent channel interference can be less, but interference to signals several channels removed will be higher. Nonlinear systems, Harris argues, not only use excessive spectra, but they also increase transmitted sideband power and so increase interference potentials for both adjacent and cochannel stations.

Harris also believes that the Kahn/Hazeltine receiver would need two integrated circuits as well as several capacitors for signal decoding because of audio phase shifting and, thereby, increase costs over the usual one-IC system.

Pioneer's Comments. Pioneer Electric Corp. of Japan also filed comments with the Federal Communications Commission and, since it is a leader among receiver manufacturers, we thought some of their remarks would be appropriate.

Pioneer favors an AM stereo signal with pilot tone and a well balanced system. But they say that no systems are free from incidental phase modulation, tuning difficulties, and microphonic noise—and these require "countermeasures" in both transmission and reception. Spectrum spreading, Pioneer found, was insignificant for practical program material. Engineers suggested a limit of 85 percent for downward amplitude modulation because of noise "breakup" in nonlinear systems, while mono modulations should remain *lower* than −100 percent. Pioneer would have receivers with a simple decoder and few production line adjustments.

These, of course, were early comments (July 19, 1979) since Pioneer at that time endorsed the Magnavox AM/PM system as the one best qualified and is now one selected by Motorola for mobile use and modifications (SX6). It now fully endorses Motorola after a 4000-mile test analysis.

Ford Motor Co. This giant automobile manufacturer also endorsed either/or Magnavox or Belar as being the simpler and less expensive systems to integrate while claiming that, without custom ICs, Motorola, Harris, and Kahn would be too complex for present radio designs. Motorola, of course, has done exactly this, and Kahn/Hazeltine will probably follow. What a difference a year or two makes as manufacturers respond to evident requirements.

General Electric. This very large U.S. marketer of AM/FM radios wants all system designs "to provide AM stereo broadcasting without perceptible degradation of service to existing monophonic receivers. It, too, in early comments endorsed Magnavox but really wanted the FCC to pick and choose among the various designs and mandate a single set of standards. Initially it did not like the Harris system, commended Motorola for detailed engineering analyses, and wasn't keen about Kahn/Hazeltine, saying receiver design was both complex and expensive. Belar also failed to gain G.E. approval because noise factors. This, of course, was another receiver/manufacturer/ distributor viewpoint, which pretty well rounds out that portion of the critique.

Magnavox. This manufacturer and initial design front runner has since farmed out its transmitter development to Continental Electronics of Dallas, and will undoubtedly have some future units available of one description or another in coorperation with its North American Philips owner. The U.S. Parent company, of course, is Philips of the Netherlands. Meanwhile, we've heard little from Magnavox since the FCC reversed itself and decided to authorize no particular manufacturer for the AM stereo service. Philips and Sony (Japan) are known to have a close working relationship, so something definite should be heard from those two as soon as AM stereo gains a little more ground. Sony already has a multi-chip decoder, not yet fully automatic, but is switch-operated for Kahn/ Hazeltine. Latest receiver and detector information will be found detailed in Chapter 5.

HARRIS ADOPTS MOTOROLA SYSTEM

On December 17, 1984, the Broadcast Group of Harris Corp. announced a licensing agreement with Motorola, Inc. to manufacturer and market AM stereo exciters and monitors using Motorola's C-QUAM AM Stereo system. Type acceptance for the Harris STX-1B C-QUAM AM Stereo Exciter has already been filed with the FCC and prompt approval is expected. Current Harris system users—and they number some 200—will now be offered a C-QUAM exciter modification kit and modulation monitor modification program. Harris says that if all its stations convert to C-QUAM, almost 400 AM stereo stations will be on the air from coast to coast broadcasting the Motorola system.

Chapter 4
AM Stereo Transmitters

While we've rather exhaustively discussed the various initial AM stereo systems and what they represent, a definitive description of transmitters has not, until now, been undertaken in detail; therefore, we shall proceed with actual hardware illustrations and text as taken directly from existing factory manuals and glossy prints. Along with integrated circuits and a careful look at some of the available receivers in Chapter 5, this will round out our presentation of AM stereo systems in the United States as they now exist. As you are by now aware, however, one or more of these four remaining systems could fold, and/or new ones could take their place under the FCC's open market policy.

About all a new entry should need would be a type-accepted transmitter, some reasonable expectation of financial success, and perhaps a loaded fairy godmother as a sponsor. AM stereo is not expected to blossom overnight, but in a long, steady haul it should give FM stereo more than considerable competition, especially in mobile radios. So from these viewpoints, let's proceed to examine several interesting transmitters, and their hardware and circuitry. We'll begin with Motorola.

MOTOROLA'S TRANSMITTER

Compatible Quadrature Modulation (C-QUAM) basically produces the sum and difference of the left and right stereo channels and modulates the broadcast transmitter with them. During the modulation process a 25 Hz pilot tone is added to the L−R stereo channel so the receiver will know that C-QUAM is being broadcast.

There are two parts to this stereo addition; one is called the stereo exciter and the other a monitor. As you can see from the basic block diagram (Fig. 4-1), the exciter does all the work while the monitor looks at the results. Our emphasis, therefore, remains on the exciter, its operation, and circuits. The actual broadcast transmitter remains in place (subject to possible modifications), as does the antenna and some studio controls, depending on their stereo suitability, age, and utility. A first class conversion would probably necessitate considerable

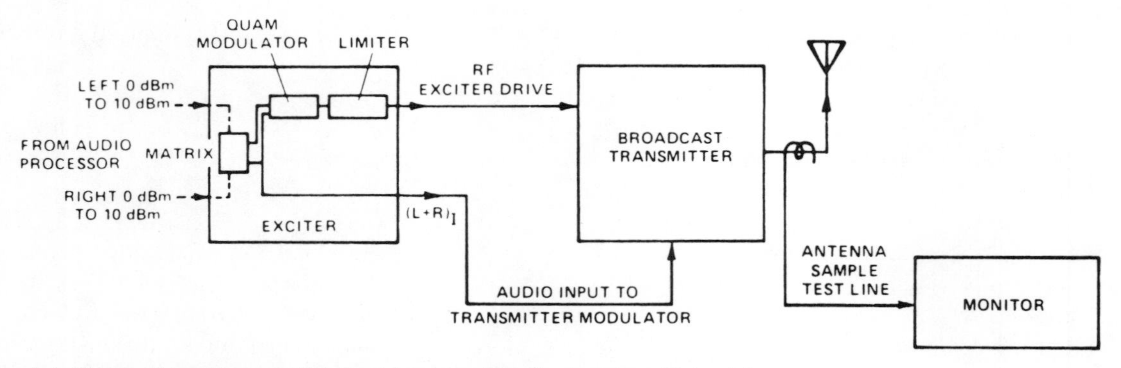

Fig. 4-1. Simple block diagram of Motorola's exciter/monitor (courtesy Motorola).

changes from microphones to the actual transmitter if it's an anitique. The exciter and monitor appear in Fig. 4-2.

C-QUAM Stereo Exciter

Specifications for this unit call for a minimum stereo separation of 45 dB (100 Hz to 5 kHz); L/R frequency response of 100 Hz to 10 kHz, ±1 dB, or to 15 kHz, ±2 dB; harmonic distortion between 0.25-0.5 percent at 85 percent modulation; and rf output adjustable to 5 watts into 50 ohms. Also adjustable are L+R, audio input and phase equalization. Stereo-monaural may be switched and indicated by front panel LED. Delays and equalization are duplicated to permit exciter operation with either two separate transmitters or two significantly different antenna systems. There's also a sample transmitter output located on the exciter's rear for diagnostics and exciter-transmitter output comparisons.

The exciter consists of a series of plug-in cards containing printed circuits boards on which are mounted the various amplifiers, oscillators, equalizers, and passive components. There's a day audio equalizer matrix card, exciter encoder circuit card, exciter L+R amplifier sample transmitter cir-

Fig. 4-2. Composite photograph of Motorola's AM stereo exciter/monitor (courtesy Motorlola).

cuit card, rf amplifier card, exciter audio night processor card, interface card, bulk delay circuit card, and an exciter power supply card.

The C-QUAM encoder is diagrammed in Fig. 4-3, where you will notice the I and Q modulators, their oscillator, summer, and limiter. These modulators are precisely 90 degrees out of phase because of inputs from the divide-by-four Johnson counter, which reduces the oscillator input by that amount and then generates digital signals exactly 90-degrees out-of-phase for the two modulators. L+R proceeds directly to the in-phase I modulator, and L−R input goes to the Q modulator which is Johnson-countered out of phase. There is also an L+R path to the transmitter audio input.

I and Q modulated signals, along with the carrier, reach the summing network (denoted by Sigma) together, and this output is pure quadrature AM stereo information. Everything then passes through a limiter to strip AM components from the signal, leaving phase-modulation Q sidebands. This is the information that finally reaches the broadcast transmitter *in place of* the usual AM transmit crystal oscillator. The L+R audio input, of course, travels by a lower path and reaches the transmitter also.

When the transmitted signal is strictly L+R, pure AM consists of I sidebands only. When the carrier momentarily approaches zero—as in 100 percent negative modulation—envelope detectors go to zero, the I demodulator does likewise, as does the Q demodulator output. So there will be no noise "popping" from the stereo channel under these conditions.

Following a phase equalizer and the matrix, the 25 Hz pilot tone is added before quadrature modulation. This phase equalizer compensates for

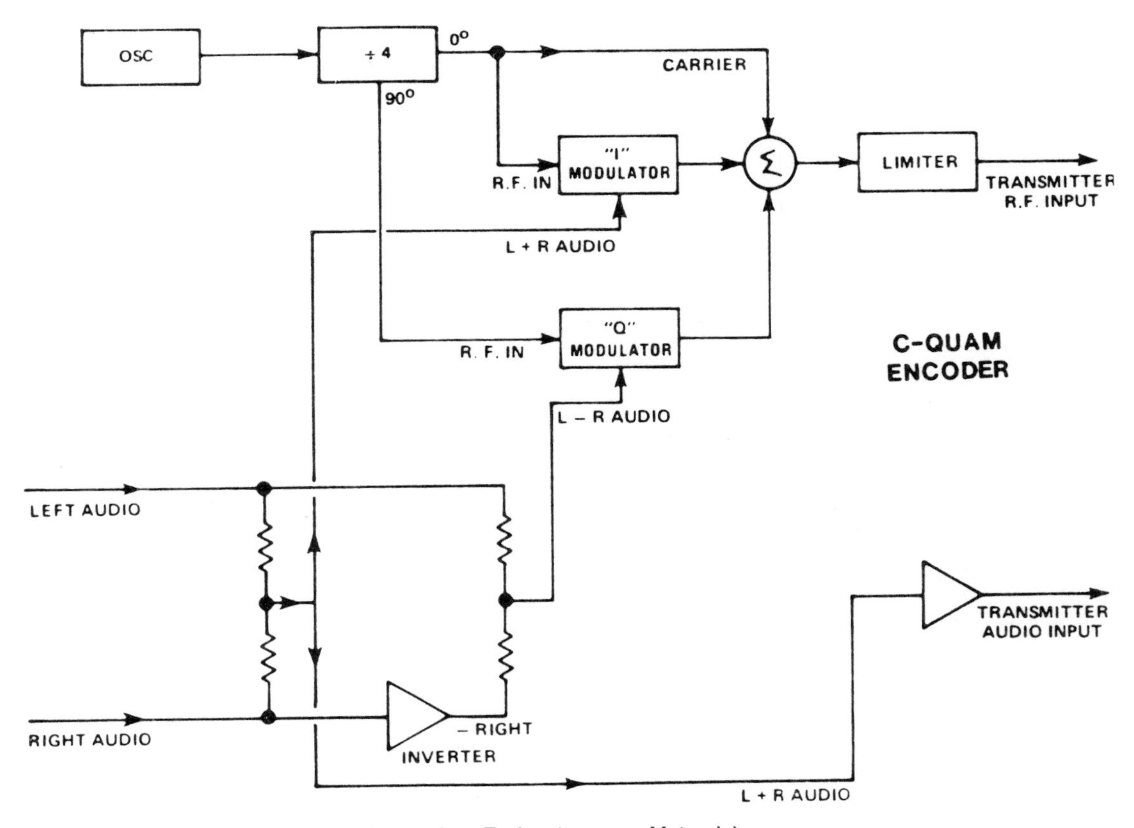

Fig. 4-3. Motorola's C-QUAM Encoder Transmitter Exciter (courtesy Motorola).

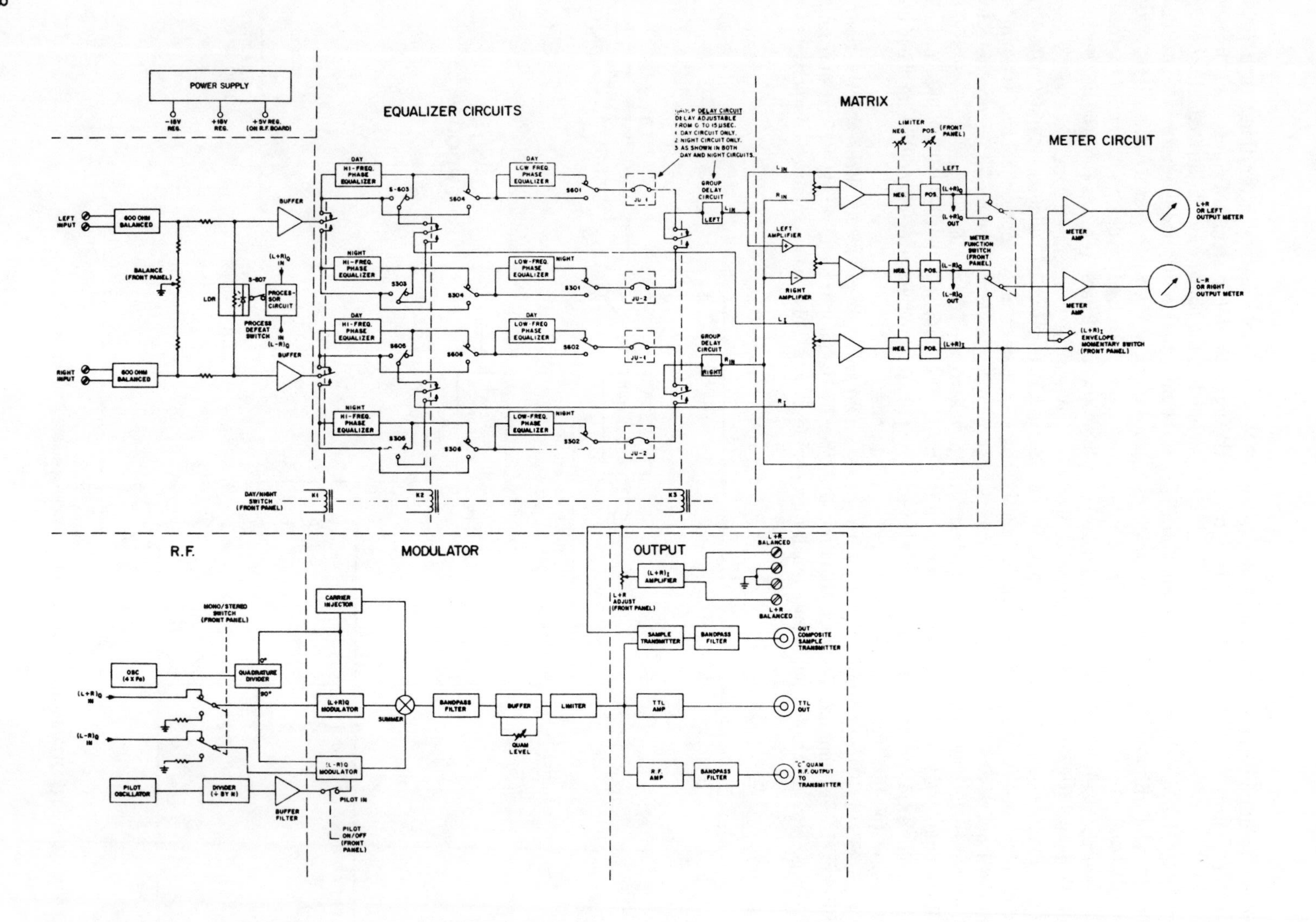

Fig. 4-4. Complete Motorola exciter block diagram showing details of the equalizer, matrix and modulator subsystems (courtesy Motorola).

differences in amplitude and phase between audio and rf to maintain the necessary wide bandwidth separation.

A complete block diagram of the entire exciter is illustrated in Fig. 4-4. Here you see in detail the power supply and its three regulated outputs, right and left balanced inputs and their buffers, the day/night phase equalizers and switching arrangements, the rf, modulator, and outputs, and finally the three-way matrix and its follow-up meter circuit. About the only portions of the exciter not previously explained are the night/day low and high frequency equalizers. These have been designed expressly for stereo stations having different day/night transmitters, coverage areas, or power reductions. What the day/night options offer is a switchable, preset phase and group delay between exciter and transmitter to take care of two divergent conditions. If discrete adjustments are required, these can be made individually on two plug-in cards to accommodate the various phase of delay situations that may arise among the many field transmissions systems.

Motorola's Monitor

Monitor main specifications include: input level, 1-10V rms; modulation meters switchable Left, Right (L=R), (L−R), or +/−; attenuation 0-50 dB; meter range 0 to 140 percent (−20 to +3 dB); peak modulation indicators, (L+R group), −100 percent indicator flashes at −99 percent, +125 percent indicator flashes above 124 percent adjustable peak indicator from 30 to 150 percent; (L−R group), negative limit flashes at 83.67 degrees, flashes when modulation exceeds 99 percent, peak flasher adjustable from 30 to 125 percent; and on the rear you will find output BNC connectors for L+R/L−R remote flashers and meters, left/right audio 600 ohms balanced/unbalanced, 25 Hz pilot tone, envelope detector, in-phase detector, and L−R quadrature detector.

Operationally, there are plug-in cards for the mixer, decoder, AVC, meter, power supply, and a shielded rf attenuator, which are inserted in the metal mainframe housing.

Since the Motorola system basically phase-modulates the broadcast transmitter while routing L+R to the transmitter's audio input, a decoder-preferred design must duplicate, to a large extent, actions of the encoder but inversely so that left and right intelligence may be recovered (Fig. 4-5). This is done with an envelope detector and an I detector.

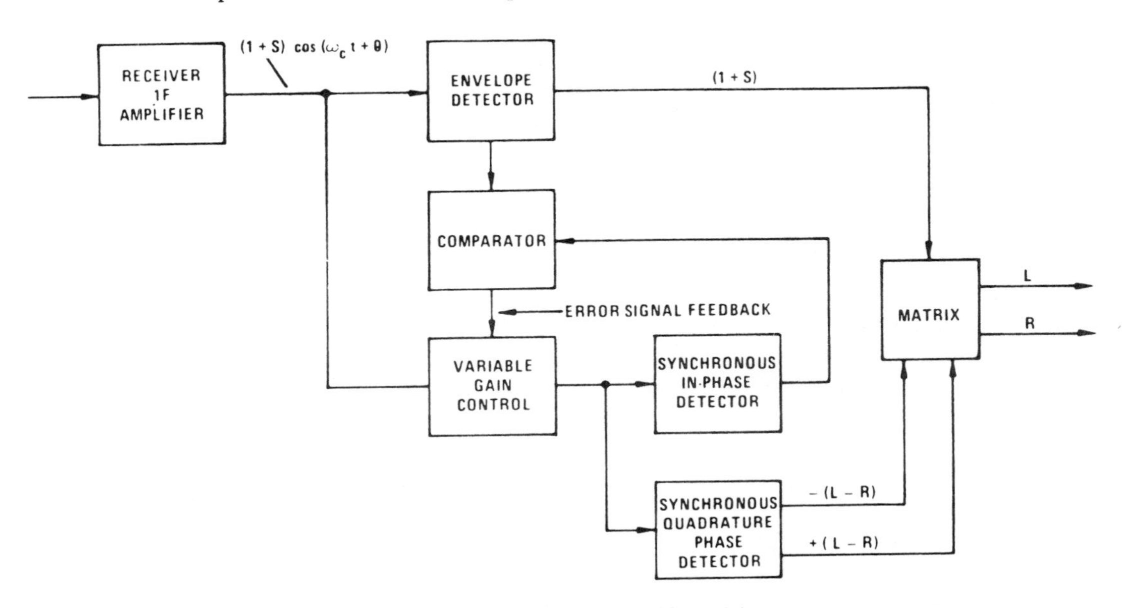

Fig. 4-5. Envelope and I detectors recover AM stereo audio (courtesy Motorola).

The two are compared and resulting error signals gain-modulate the I and Q demodulators. When monaural is received, it's in the form of pure AM or I sidebands and passes through an envelope detector. But when AM and PM are transmitted, the input is phase-shifted to the I demodulator. The envelope detector is then compared with the I demodulator and if there's an error, the detector level is raised, making inputs to the I and Q demodulators appear as pure quadrature. The audio output then delivers a "perfect" L−R signal, and the demodulated output is combined with that of the envelope detector to produce left and right audio.

So, except for the various manual and automatic metering functions, that's the monitor. The functional block diagram in Fig. 4-6 sheds a little more light on the general operation and substantially illustrates the panel switches, outputs, associated detector, buffer outputs, LED modulation, and limit indicators. Attenuated rf input begins at the upper left-hand portion of the diagram, and most outputs are on the right.

KAHN/HAZELTINE

The independent sideband system (ISB) proposed by Kahn/Hazeltine places lower sideband(s) in the left stereo channel and upper sideband(s) in the right stereo channel, making it "essentially a frequency separation system" versus phase separation for most competing systems. Leonard Kahn believes this produces a considerable difference in performance under adverse conditions "such as skywave/groundwave interference, sensitivity of transmitting and receiving antennas, and possible use of advanced receiver technology such as asymmetrical sideband selectivity."

The Kahn/Hazeltine system, however, does, in itself, begin with actual phase separations of ±45 degrees, but once the independent sinewaves are generated, receivers tuned to upper and lower sidebands react to frequency and not phase. A block diagram of the exciter-generator is shown in Fig. 4-7.

Right and left audio inputs are summed and then applied to a "phase difference," constant amplitude arrangement to excite a standard AM monaural transmitter (lower diagram portion). L and R information then pass to another phase difference circuit where sum and difference audio signals emerge in quadrature and both enter a phase modulator. Thereafter they are frequency doubled, amplified in a level squarer—voltage gain amplifier circuit—summed with a 15 Hz pilot signal, and then time delayed for proper transmitter match. The signal is then applied to a crystal-controlled phase modulator, and the carrier oscillator "translates" this information up to the frequency of the transmitter.

Unfortunately, this is all the information provided by Mr. Kahn, and although we would like to offer considerably more detail, further transmitter or monitor data are considered "company confidential." Another highly simplified block diagram of a Kahn receiver appears in Fig. 4-8, showing an rf/IF block, along with the L+R information being demodulated in the envelope detector, some feedback to the inverse modulator, and stereo information being detected via the carrier track in a synchronous demodulator. A phase difference network and matrix then separate the two for final receiver stereo amplification. Once again we apologize for very restricted information, but nothing else was released. The system concept does appear simple, but execution could be somewhat complex.

Mr. Kahn, however, does deplore certain features of other systems he thinks detract from good stereo enjoyment; one of them being a popping sound that appears under certain conditions in the Magnavox sets. That company quickly replies to this difficulty and discusses a little more for the record, which makes interesting reading.

Noise Problems. "Noise pops" in receivers, Magnavox says, occur when the stereo signal "vanishes" during −100 percent modulation peaks, and, "it is impossible for *any* AM stereo system to operate flawlessly." Tradeoffs, says Magnavox, have to be made between a "degraded" signal-to-noise ratio or other distortion when a deep modulation peaks appear during stereo reception. And the company recommends a negative modulation peak limitation of 95 percent to avoid this problem; but would abandon the suggestion "if all of the new AM

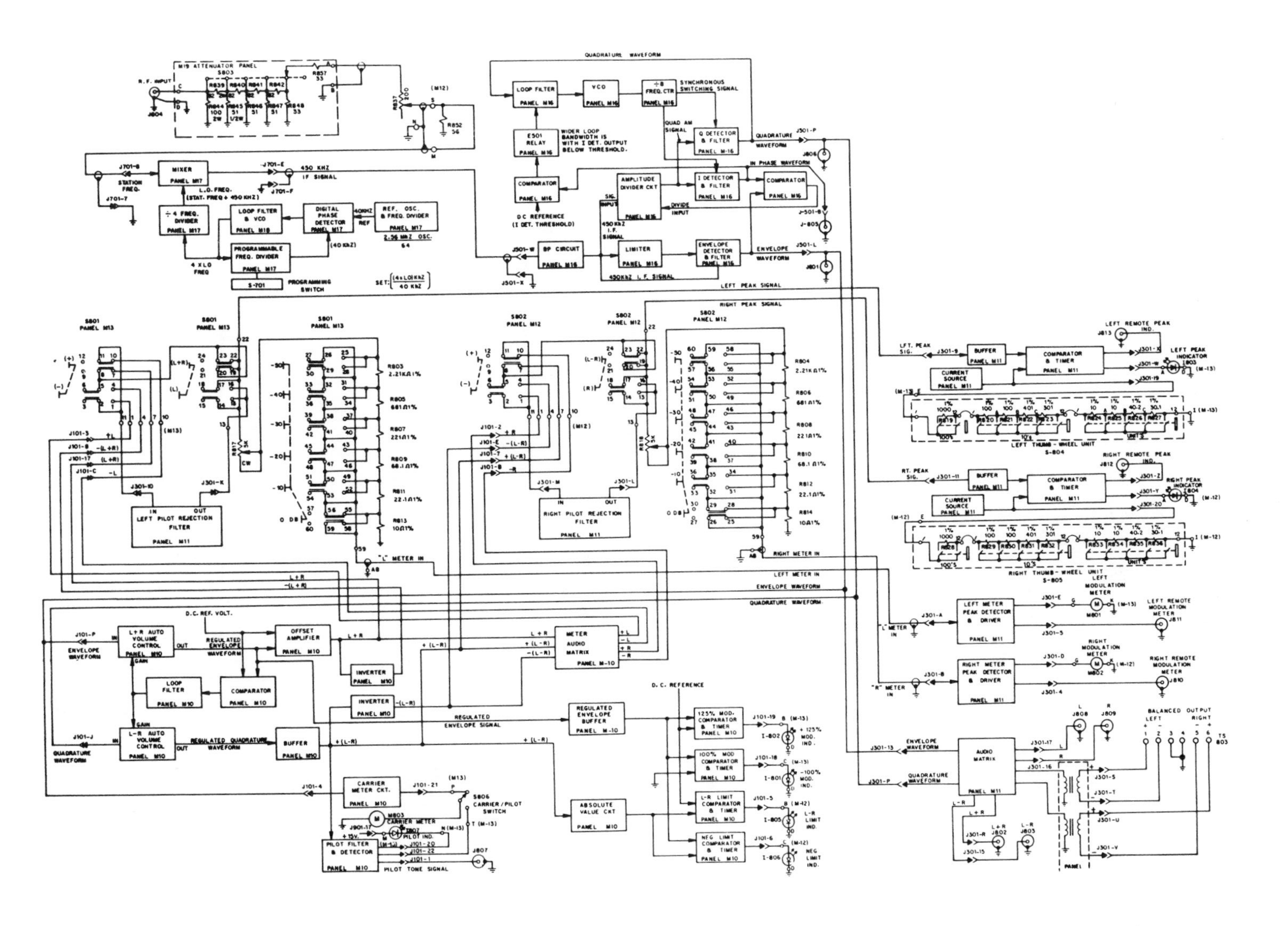

Fig. 4-6. Motorola's monitor functional block diagram (courtesy Motorola).

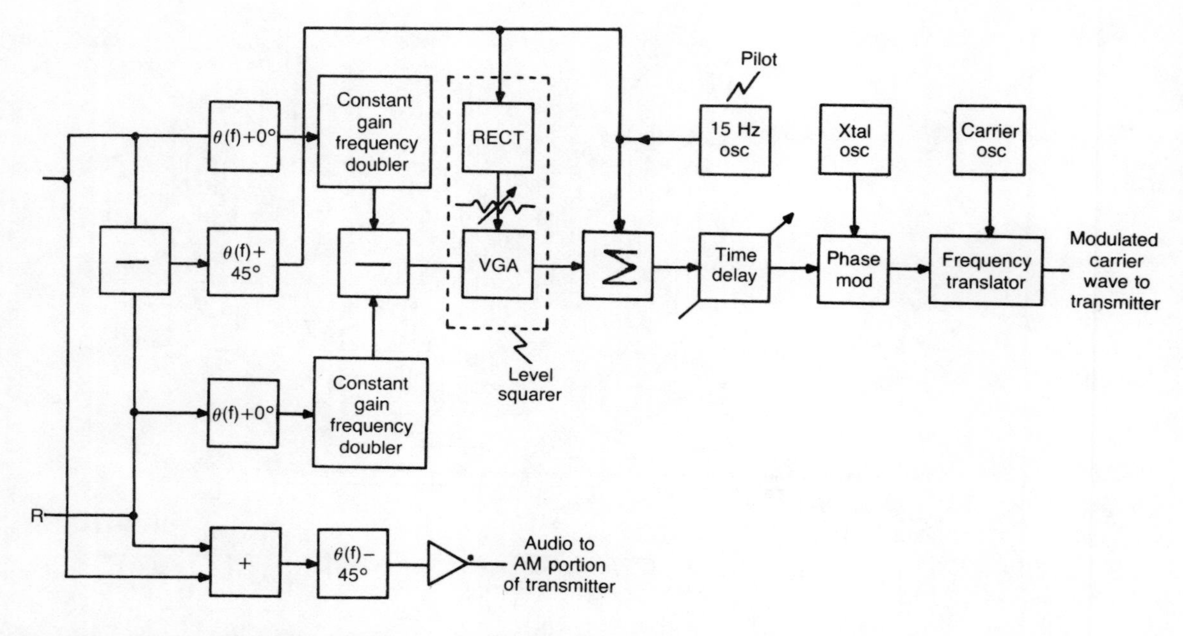

Fig. 4-7. The Kahn/Hazeltine Simplified exciter block diagram (courtesy Kahn/Hazeltine).

stereo decoding techniques which evolve are commercially viable in operation at −100 percent negative peak modulation."

The company then explores other methods and directions of relieving this possibly annoying situation. Elimination of "some of the sideband energy" before detection will reduce modulation depth and should eliminate almost all −100 percent noise bursts—but Magnavox cheerfully admits this technique isn't foolproof and does produce a low fidelity receiver. Another method, they say, might be to enhance the carrier relative to the sidebands. IFs may be modified to permit more or less fidelity, but could contain a small center passband peak where the carrier would have been placed. Or, other techniques such as a tracking phase-locked loop could be used depending, of course, on how sophisticated you'd like to become and the ultimate

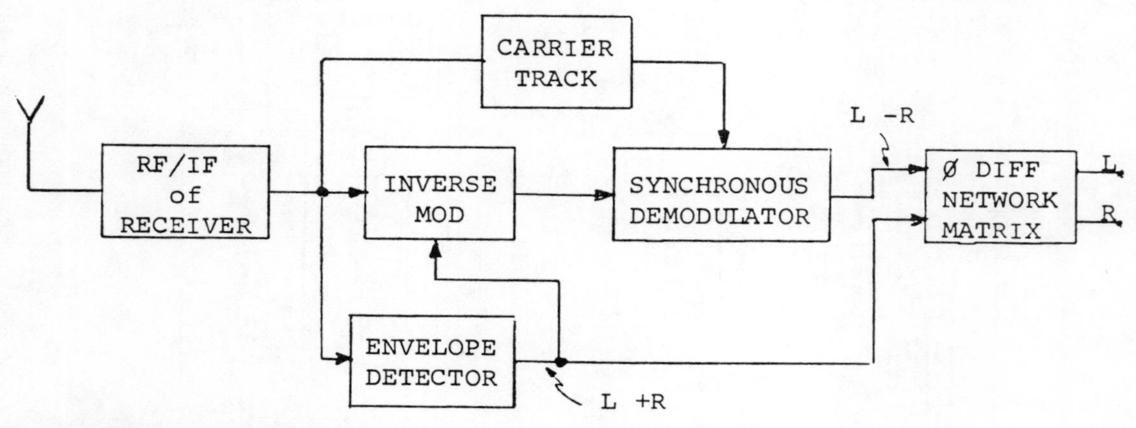

Fig. 4-8. Kahn/Hazeltine receiver block diagram.

cost to the consumer.

There's also the inverse modulator which is known to Leonard Kahn and, in at least one application, has been patented by Motorola, according to Magnavox. But, again, Magnavox engineers warn that *all* protection techniques require tradeoffs between noise bursts and "distortion effects." Finally, Magnavox makes the claim that the Motorola/Magnavox system approaches are "essentially" the same "except that Magnavox's is significantly less complex and therefore less costly to implement, and has a better pilot tone design." But Magnavox does state that with a proper "cos ϕ" corrector, National Semiconductor's LM1981 integrated circuit used in the Magnavox receivers will also decode the Motorola signal. Of course, this IC does *not* decode the pilot signal which is essential to stereo recognition.

One interesting aspect of the Magnavox report is that it had already discussed Kahn exciter conversion possibilities with the Kahn/Hazeltine group and, subsequently, had actually modified one of these units to broadcast the Magnavox system. Magnavox, however, makes no claim that the "commonality" between the two systems nor the conversion and decoding techniques were easy. Further, since detailed information on the Kahn exciter remains confidential, there will be no overall release of specific information. The company did say there were modifications made to the low pass filter on the Kahn exciter's card 1, added test formats on card 2, adjusted delay characteristics on card 5, and a change or two on rf power amplifier output, card 10. Magnavox engineers also say they have "available" master oscillator modules for many of the stations known to own Kahn exciters." Several block diagrams of these cards are included in the modification explanation but will not be shown here since it involves experiments rather than a full-fledged commercial product. This may suggest that conversion of one system to another may not be that difficult—but we really don't know the full story.

Magnavox is also pushing the use of its 5 Hz pilot tone as a means of transmitting "low rate" digital data. They say there would be no interference with audio performance and would permit the sending of virtually error-free data. The company has already breadboarded a circuit to replace the 5 Hz pilot tone with a 5 Hz square wave, "equivalent to 5 bps Manchester-coded baseband." Differential encoding of "NRZ data" is suggested to prevent polarity ambiguities.

MAGNAVOX

This system—first accepted by the Federal Communications Commission and then set aside—has not really changed in principle, but transmitter manufacturing has now been turned over to Continental Electronics Manufacturing Co. of Dallas, Texas and exciters and montiors are available within several weeks of order receipt. Continental is an engineering firm specializing in design, development, and production of low, medium, and high power radio frequency transmitters and offers both AM and FM transmitters.

The Magnavox/Continental AM stereo exciter is a Type 302A (Fig. 4-9) for which the instruction manual was printed September 5, 1983. Briefly, the Magnavox PMX system is a mixture of amplitude and phase modulation with left and right stereo matrixed and broadcast. Phase modulation peak-deviates at one radian, accompanied by a 5 Hz pilot stereo ID modulating the carrier up to 20 Hz deviation. The L+R information amplitude modulates the carrier and is the mono signal. Exciter rf operates above the broadcast band and is heterodyned down for transmission.

The 302A Exciter

The exciter is a self-contained, rack-mounted package of electronics that delivers frequencies between 50 Hz to 10 kHz (0.5 dB) with less than 1 percent harmonic distortion. Noise appears at more than 60 dB below +10 dBm at 1 radian phase deviation, and stereo separation is specified at greater than 35 dB from 100 Hz to 5 kHz, decreasing by 6 dB/octave below 100 Hz and above 5 kHz. The external pilot input amounts to 1 V p-p into 1 kilohm; the rf output, 2 W into 50 ohms (adjustable), and frequency stability remains within 5 Hz over 0 to 50 degrees C. Rackmount weight is 32

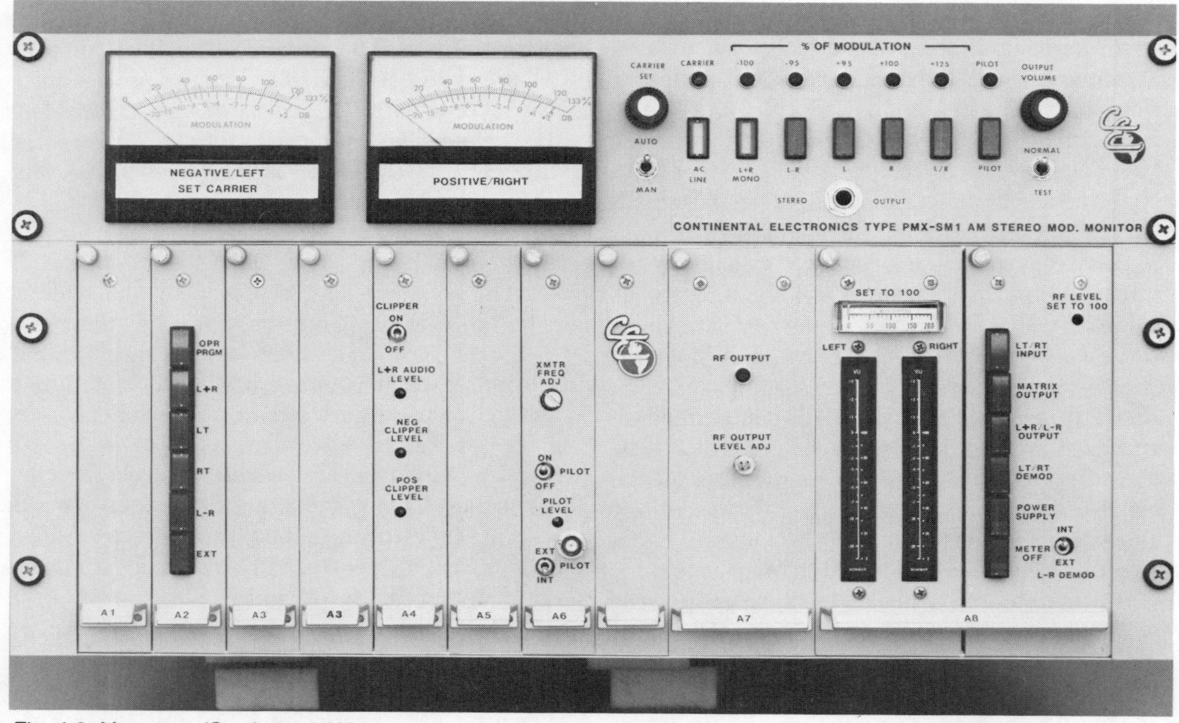

Fig. 4-9. Magnavox/Continental AM stereo exciter and monitor combination (courtesy Continental Electronics).

lbs., including eight plug-in cards and a power supply. Their descriptions follow:

Audio Input (A1) has transformerless balanced input amplifiers to receive left and right stereo signals with terminating impedances of either 600 or 20 K ohms. There's also a switch-selected low pass filter with cutoff at 20 kHz.

Matrix Generator (A2.) Audio signals from A1 are matrixed into L+R and L−R equal amplitude outputs to drive envelope and phase modulators. A selector switch provides various test modes.

Phase Equalization (A3). Receives L+R and L−R, has 31 μsec phase delay in 1 μsec steps for maximum stereo separation-equalization for phase and envelope modulation channels.

Audio Output Amplifier (A4). L+R from A3 is amplified and produces high level balanced audio output of approximately +20 dBm.

Phase Modulator (A5) processes L−R and rf from carrier frequency generator. Rf is amplified and shaped before entering a 3-stage varactor phase modulator, then amplified again by the rf power assembly. L−R is amplified and then drives the modulation input of the 3-stage varactor modulator. A clipping amplifier limits L−R above 120 percent modulation.

Carrier Frequency and Pilot Tone Generator (A6). Fundamental carrier generated by a crystal accurate to ±100 Hz while a 5 Hz pilot oscillator sources from a function generator. The pilot adjusts for peak deviation of ±20 Hz and may be switched on or off.

Rf Output Amplifier (A7). This, the final card in the exciter chain, receives rf from the phase modulator and amplifies it to drive the transmitter rf input. A one-shot multivibrator for pulse shaping is available for maximum rf amplifier efficiency.

Power Supplies. PS1 and PS2 respectively, furnish ±24V for each card and ± 5V for the VU meters. Exciter input requires less than 25 watts.

Demodulator & Monitor (A8) provides an independent plug-in AM stereo demodulator and monitor for audio levels in the exciter and demodulates transmitter AM and PM outputs for transmitter adjustments and stereo monitoring.

PMX-SM1 Modulation Monitor

The PMX/SM1 is a companion unit to the 302A exciter just described that decodes and displays AM stereo broadcast signals to analyze their operating conditions. Requiring but a single rf input, the monitor offers internal metering for L+R and L−R, left and right channels, and pilot tone transmissions. Rf inputs are adjustable, stereo volume amplitudes may be varied, and there's also an earphone connector in addition to peak modulation indicators and an adjustable indicator for negative envelope peak modulation. Outputs are a balanced 600-ohm left and right channel audio, an unbalanced +10 dBm 600-ohm left and right channel audio, an AM SCA output, and a test output.

The stereo detector module (1A1) is a precision decoder consisting of four circuit cards: an AM envelope detector with auto rf leveling; an angular modulation detector and pilot tone recovery system; an audio matrix and filter for L−R, plus left and right channels; and a pilot tone filter and high speed AM SCA/pilot tone detector.

Power Supply (1A2) furnishes ±37.3-volt full wave dc potentials which supply two voltage regulators at ±15 V for the modulation monitor. Current limiting helps prevent power failure in the event of equipment short circuits or excessive current drain.

Meter Driver (1A3) receives audio signals for measurement from the front panel selectors. Amplification, buffering and peak detection occur before current enters each meter which has electrical zero set, calibration adjust, and full scale deflection. One meter also has polarity inversion.

Calibration Technique

During NAMSRC investigations, Magnavox discovered that when excited by a single frequency audio source quadrature signal, the Magnavox, Motorola, and Belar systems produced a common result: a nulling effect for the first order sideband on but one side of the carrier. After careful investigation, the engineers found this was a reliable and "sensitive" means of system calibration.

They discovered that when "carefully" (probably means precisely) formed L+R and L−R signals are developed and adjusted for exactly the proper modulation percentage and there is a system-specified zero time (delay) between L+R and L−R, "a nearly perfect null in one of the first order sidebands will be observed." When the right channel lags the left by 90 degrees the upper sideband nulls, and when the left channel lags the right by 90° degrees, the lower sideband nulls. Therefore, says Magnavox, with other adjustments already made, "modulation of the system with a single-frequency quadrature audio signal allows identification of the proper phasing structure in the entire audio system, permits precise adjustment of carrier modulation conditions for L+R and L−R , and provides a precise technique for obtaining the proper time delay in those two signal components."

A diagram that's useful for such calibration is included as Fig. 4-10. It is also said that a 0.1 dB error adjustment of L+R and L−R ratio level will reduce the null magnitude from −130 dB to −48.3 dB, and a 1 dB error moves the carrier null to only −28.8 dB, etc. On the diagram, the horizontal axis is calibrated in both dB and voltage ratios for L+R and L−R, and the vertical axis is calibrated in the dB null magnitude of the undesired sideband "relative to the carrier."

Magnavox says there are six parameters when computing the calibration null: left and right signal magnitudes, signal phase offset, L+R and L−R magnitudes, and their phase offsets. In using signal injection test equipment, engineers say there are some problems in obtaining low distortion quadrature audio sources covering a wide frequency range. So they built one and use it "regularly" with a spectrum analyzer to check their equipment. The signal is injected directly into a transmitter matrix which produces L+R and L−R. An initial adjustment of the L+R and L−R modulation amplitude will usually ensure virtually correct system operation. If necessary, the L+R/L−R delay can then be adjusted to relative zero by observing the null "be-

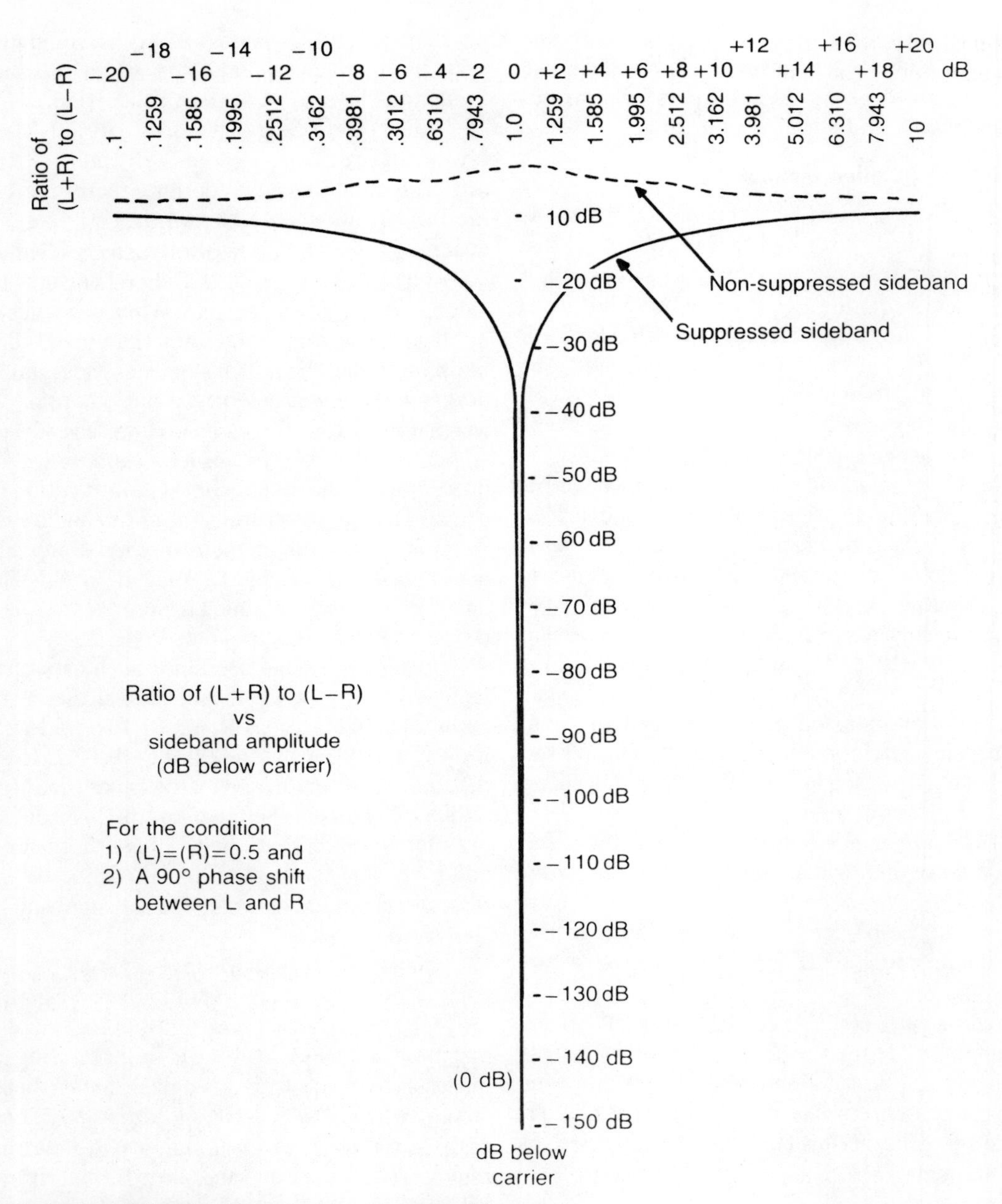

Fig. 4-10. Carrier sideband nulling adjustment (courtesy Magnavox).

havior" as a function of frequency. When this has been done, final adjustments are made to the L+R and L−R amplitude ratio, and the null condition is fully accomplished.

Undoubtedly there are other equally effective methods of tuning exciters, but this one seems to have considerable merit because of its simplicity and apparent accuracy. And since it was developed several years ago, satisfactory equipment is probably already on the market, offered, perhaps, by

Boonton Electronics and others specializing in audio generators. You may find it worth looking into.

HARRIS

With its STX-1A exciter fixed 90-degree quadrature modulation now fully approved by the Federal Communications and probably 100 or more broadcasters using the system on the air, Harris is moving quickly to consolidate its AM stereo position throughout the U.S. and foreign countries as rapidly as possible. And, in cooperation with Sansui, the company plans to pursue its aggressive marketing program, even to the point of offering specific modifications for conventional (better quality) AM mono radios so that you can receive the Harris system. This aspect will be covered in Chapter 5 since it is not strictly in the exciter/monitor category.

Although Harris admits to "slight" distortion when its quadrature system is received on mono radio envelope detectors, the company says it's almost all second harmonic and "seldom audible and not objectionable." But with synchronous detection— which will become available in the 4-chip decoders—the Harris system would be in its element and reproduce audio frequency responses to 15 kHz with 1 percent or less harmonic distortion.

The Harris system, of course, is a linear system where inputs applied together represent the sum of several individual inputs introduced separately. Harris claims that, in nonlinear systems, by distorting in-phase and quadrature phase, AM stereo signal elements can compensate for envelope detection distortions. But AM radio filtering is usually narrowband, and nonlinear AM stereo information will also generate distortion components in both mono *and* stereo receivers. Both harmonic and intermodulation distortion can appear.

Harmonic distortion appears in the form of shortened or phase-distorted high order sidebands. Tone intermodulation distortion occurs in nonlinear systems when the inital sidebands are outside their filter envelope and higher ordered sidebands are inside. This results in "trash" appearing among low frequencies and can be signficant, considering the usual pre-emphasis of AM higher modulating signals, often by as much as 20-30 dB. It is objectionable.

Associate Principal Engineer David Hershberger of Harris Broadcast Division points out that linear quadrature modulation has the same rf bandwidth as monophonic AM, there's no spectrum truncation distortion or problem with preemphasis, and is fully compatible with synchronous detection. He adds that synchronous detectors will not distort because of receiver detuning, will have no fading or AM multipath distortion, will welcome directional transmissions, will not distort on modulation overshoots, will have no crossmodulation, and will exhibit some reduction of impulse noise.

Compatibility with existing monophonic receivers, Mr. Hershberger continues, remains a very complex stereo requirement and there are many types of distortion in audio systems that are relevant. Fortunately, low order even (2nd, 4th, 6th) harmonics are the least objectionable while odd order harmonics (3rd, 5th, 7th, etc.) are disturbing. Second harmonics can even be made to appear as "increased treble boost," and develop no distortion at all. He further predicts that wide use of linear quadrature modulation will produce long overdue improvements to AM radios that will result in wider IF bandwidths, synchronous detection, less distortion, and greater frequency response.

Orban Associates supports Mr. Hershberger's assessments by saying that competing nonlinear systems depend on "infinite bandwidth, phase linear" transmissions for complete cancellation of standard envelope detector distortion. And IF filters do produce receiver distortion regardless of the detector used. By contrast, Orban states that Harris can offer audio without distortion with a linear synchronous detector, "regardless of IF bandwidth" provided, of course, the receiver is properly designed in other respects. That, of course, would be a decided plus in instances where interference from other stations which commonly occurs during nighttime operation—would require greater receiver selectivity. So it's no wonder that

Harris is strongly endorsing synchronous IC detectors for amplitude modulated stereo.

The Harris Stereo Exciter

The STX-1A rack mounted, 30 lb. bay of electronics (Fig. 4-11) has synthesized operation between 535 kHz and 1710 kHz in 1 kHz increments. Rf output is adjustable from 0 to 0.4 watt (0-14 volts) into 50 ohms at operating temperatures of 0°C to +50°C up to 13,000 feet. Required audio inputs are +10 dBm for 100 percent modulation at 400 Hz, with peak limiting beginning at 0 dBm. Audio input to the transmitter can be adjusted to +6 dBm into a 600-ohm balanced load. Left or right channels may be modulated ±80 percent, L+R−100 to +135 percent; L−R ±100 percent, with minimum stereo separation pegged at 30 dB between 400 Hz and 5 kHz, or 25 dB from 5 kHz to 15 kHz. Audio output S/N is greater than 55 dB and maximum mono or stereo distortion is specified at 1 percent or less.

Front panel controls and readouts are clustered around two meters and two multifunction switches. They include: multimeter, modulation meter and switch, rf voltage, ±15 V bus level, peak limiting gain reduction, feedback, static carrier phase, left and right channel modulation levels and main channel level, Q stereo level, pilot injection level, envelope modulation, audio level into transmitter, phase error between rf and feedback, peak or average modulation, rf level adjust, LED day and night mode lamps, ready LED, pilot LED, stereo LED, mono LED, and stereo/mono switch.

Inside the exciter you will find the following cards: audio input, separation corrector/peak limiter, quadrature generator, meter board, power supply, rf amp., audio output, and rf output. On the rear of the exciter you will find screw-fastened connections and BNC outlets for rf sample and rf output. A 4-ampere fuse protects internal circuits from current overloads. There are also provisions for remote switching, with control ports optically isolated from internal circuits.

The *basic theory of operation* is relatively simple since both in-phase and quadrature carriers are modulated alike with L+R and L−R information, respectively. Since both channels are amplitude modulated they contain only first order sidebands in mono. In stereo, left and right audio pass through a matrix, becoming L+R and L−R, and L−R under-

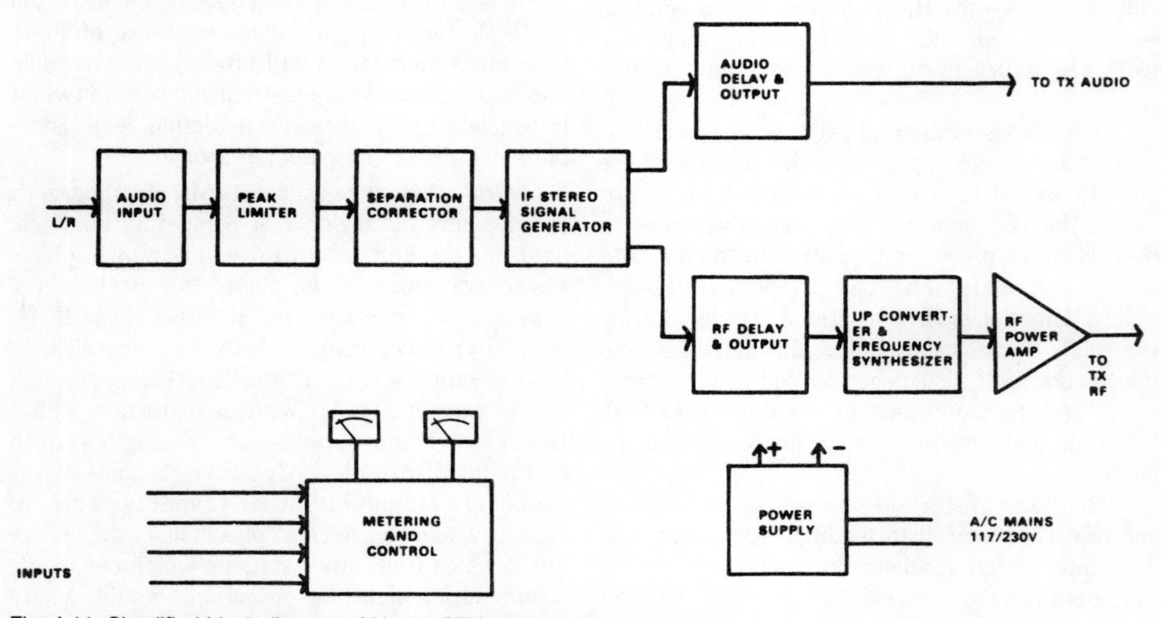

Fig. 4-11. Simplified block diagram of Harris' STX-1A exciter (courtesy Harris).

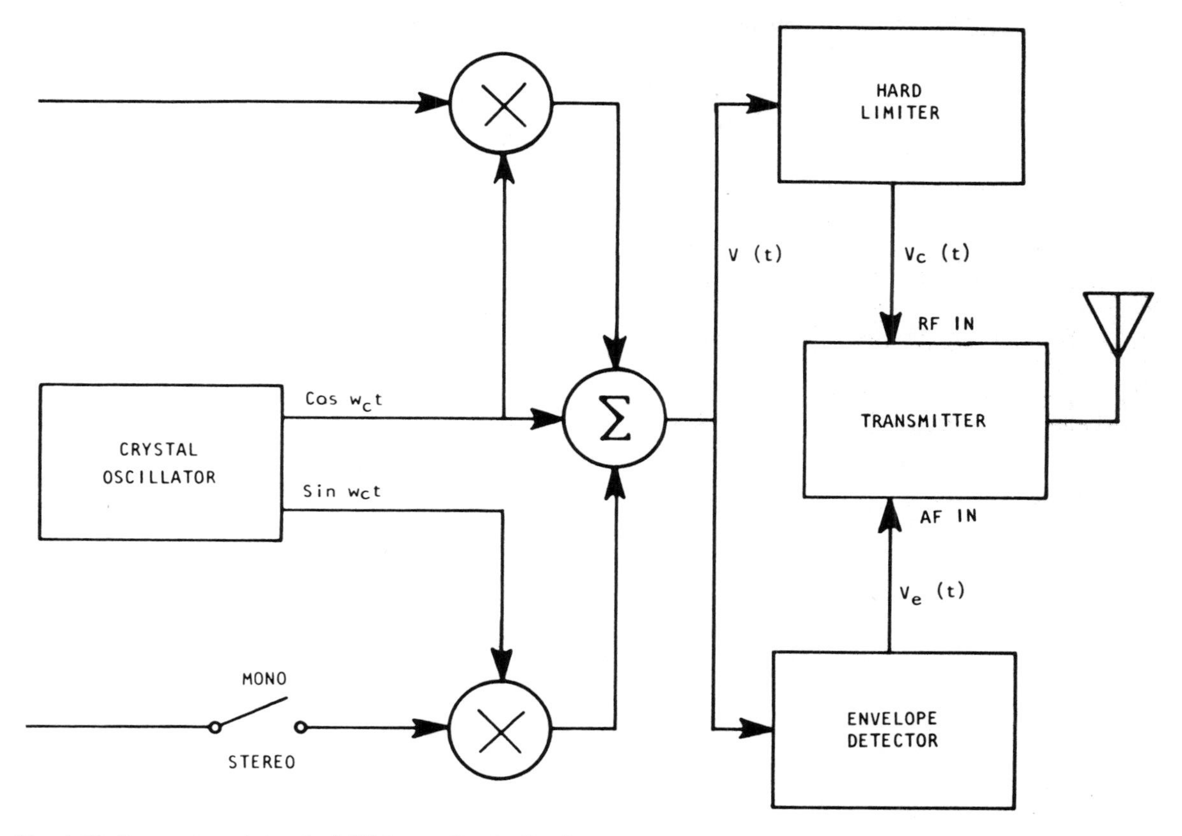

Fig. 4-12. Conversion of standard AM transmitter to Quadrature transmitter.

goes highpass filtering so that pilot signals may pass below 200 Hz. L+R and L−R then reach the quadrature generator.

In Fig. 4-12 you see a block diagram of what it takes to convert a standard AM transmitter to quadrature stereo. A crystal oscillator and phase shifter produce both in-phase and quadrature carriers so the stereo generator may output both audio and radio frequency signals. This and all other systems modulate both amplitude and carrier phase (PM). The phase portion is contained in the rf signal. Audio frequencies go to the transmitter's audio input while PM travels to a low level rf receptor in place of the transmitter's crystal oscillator, which generates the standard carrier. Both signals are then individually amplified and mixed in a modulated transmitter stage for the stereo signal. Along the way, these carriers are multiplied by I and Q audio via balanced mixers, then one is hard limited and the other envelope detected before final stereo transmission.

For good quality signals this means that all frequency components in the stereo (PM) signal have identical time delays. The audio frequency amplifier and components must also have constant time delays in addition to a flat frequency response of information originating in the stereo generator. Transmitter and antenna must have a good impedance match and be able to pass the overall frequency response of the entire signal.

Existing transmitters, of course (especially old ones) may have some real problems: neither frequency response nor phase linearity may be present; there may be low frequency rolloffs as well as high frequency attenuation; harmonic resonators and high Q interstage coupling networks can affect phase in the transmitter; and time delay differences may appear between audio and rf channels. If there is such a time delay, channel separation and distortion can become affected in additive systems and some multiplicative systems also. There's also incidental phase modulation to contend with that's the result of PM signal feedthrough via the driver-transmitter output. Here both channel separation and distortion are again affected. Another incidental phase modulation difficulty is due to transmitter nonlinear impedances in the modulator, producing phase shifts. Storage times in transistors may also vary with application of amplitude modulation, and extraneous phase modulation can develop as noise in the L−R stereo channel. This may be due to phase modulation hum, low frequency oscillator noise, or PM white noise.

Dave Hershberger and Cliff Leitch, who developed this information, say there are cures for such problems. Feedback, they say, in transmitter audio frequency chains might help, but usually open loop amplitude and phase correction in the stereo generator will minimize the problem. Then the respective subsystem is adjusted for best square wave response, and almost all transmitters will equalize for flat response to 20 kHz for additive systems or 10 kHz for multiplicative systems. Additive systems, they continue, that are compatible with monophonic radios are also compatible with modulators. Then careful neutralization and tuning aids incidental phase modulation, while extra power supply filtering reduces hum, and a new oscillator in the stereo generator often gets rid of this kind of noise.

Some redesign in transmitter outputs may also be necessary since any mono problems will be intensified in stereo. This is especially true in multiplicative systems where 40 to 60 kHz bandwidths may be required for full signal emission. All these problems should be solved before broadcasting stereo since false stereo indications and subsonic FM noise can cause real problems with lower priced sets that don't have automatic stereo-mono switching. Auto switching can be tricky too. "Dirty" mono signals shouldn't trigger stereo pilot detector systems, whatever their origins.

Finally, these engineers say that most AM transmitters can be usefully converted to stereo by adding a stereo exciter and applying relatively few circuit changes. But where transmitters are old or in poor condition, conversions may be difficult or impossible. Details should be worked out by the transmitter or exciter manufacturer beforehand so the conversion process may become fairly smooth and nondisruptive. And that, substantially, is the story of the Harris exciter. We'll now continue with the companion monitor, which tells us what the exciter is doing.

The Harris Stereo Monitor

Called the STM-1, this monitor "sees" the entire stereo AM signal; has built-in circuits eliminating the need for a spectrum analyzer during interface; features pilots frequency injection and calibration monitoring for quick system checks; and has three large meters for highly visible system analysis, thus fulfilling all FCC AM stereo requirements. A front panel switch with sideband detection permits transmitter alignment without a spectrum analyzer. The three 3.5-inch meters monitor left and right channel modulation and one becomes a multimeter. A simplified block diagram of the monitor appears in Fig. 4-13. Signal demodulation is by synchronous detection.

Briefly, what you see in this signal flow block diagram amounts to the major operations of the monitor and a few details. Amplified rf and IF enters the monitor through their respective AGC-controlled amplifiers, the synthesizer and its PLL circuit confirms and corrects frequency lock, mixing occurs for the I and Q signals, and these either proceed through additional low pass filters and amplifiers to the matrix, or phase-shifted information is switched directly into the matrix. Otherwise, L−R and L+R stereo reaches the matrix and is converted there to the usual left and right outputs. As you can see, the Q channel receives a 90-degree phase shift while the I channel has none, indicating true quadrature processing and detection.

Specifications include L and R separation between 30 and 35 dB from 400 Hz to 12.5 kHz; pilot frequency calibration ability ±0.1 Hz; monitor

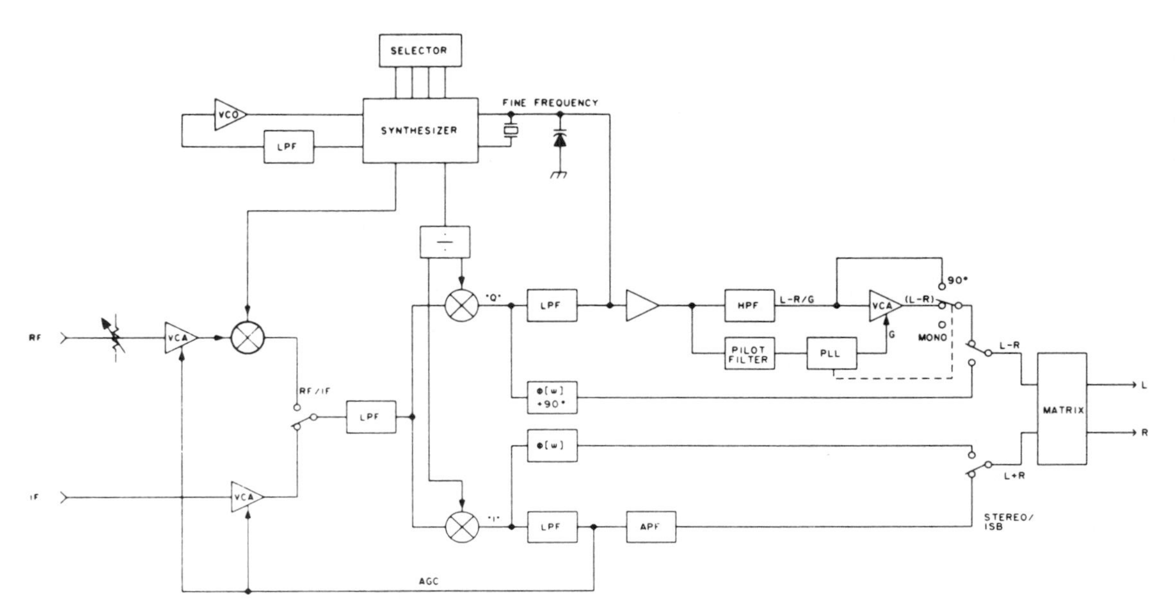

Fig. 4-13. Harris' Monitor (courtesy Harris).

modes, forced stereo, auto stereo, mono; left and right frequency response from 400 Hz to 15 kHz; I and Q outputs; head phone jack; stereo pilot LED; and rf input 1-10V rms into 1K impedance.

Waiver Granted. Harris had a little problem with single-tone distortion and the FCC's ruling in Section 73.128(b)(2). This requires that harmonic distortion by an envelope detector "not exceed five percent under specified measurement procedures." The FCC said such distortion could rise as high as 30 percent using single tone modulation "under certain limited conditions, but that listeners did not find distortion highly objectionable in subjective tests using actual program material." Industry comments confirmed this finding and several noted that Harris operates without distortion, at least as certain levels, when received by sets having synchronous detection.

The Harris problem began when the FCC found the STX-1 stereo exciter offered by the company operated "differently" on program material than it did with test tones. Originally, the STX-1 exciter was marketed as a "variable angle compatible modulation system." This was done so that the L+R and L−R phase angle separation would change, permitting better compatibility with mono receivers having envelope detectors. During stereo programs, however, "an internal circuit" locked the carrier at full 90-degree quadrature.

As a result, Harris modified its STX-1, which was temporarily removed from the market, to a fixed phase angle system, and the FCC granted the company a waiver on the offending section, saying there was "no evidence that the Harris system will cause co-channel or adjacent channel interference to other stations." The modified unit bears Model No. STX-1A and is now known as a "controlled compatibility quadrature system" since its variable frequency pilot tone is now fixed. We also understand that Harris has changed its pilot frequency to 25 Hz, the same as Motorola's, and that Harris transmissions may also be received on C-QUAM but with some distortion. (Harris has now adopted C-QUAM and is licensed by Motorola.)

BROADCAST ELECTRONICS

The newest entry into the AM stereo exciter market (as far as we can determine) is an Illinois company identified as Broadcast Electronics, Inc., a

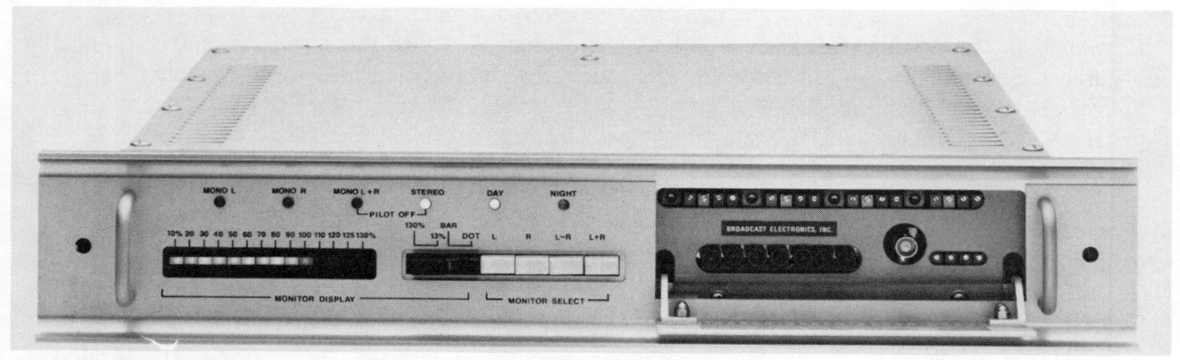

Fig. 4-14. Front view of the AX-10 AM stereo exciter by Broadcast Electronics, Inc.

manufacturer of FM equipment. Under license to Motorola for C-QUAM, the AX-10 exciter is being released to transmitting stations as "second generation C-QUAM design and for high performance AM stereo broadcasting. Offering independent, noninteracting left and right channel modulators "in an IF modulation configuration," the AX-10 is said to interface with almost any existing transmitter, yet be fully compatible with existing monophonic receivers. It also has full remote control selectivity and a built-in LED bar graph for peak reading modulation, as well as independent equalization two-transmitter or dual antenna pattern operation. A photograph of the unit is shown in Fig. 4-14. An MA-1 AM stereo modulation monitor is also available. Specifications for the unit are as follows:

Rf power output: 0.15-10 watts variable rms into 50 ohms.

Frequency range: 522-1620 kHz.

L+R audio out: adjustable −10 to +26 dBm into 600 ohms.

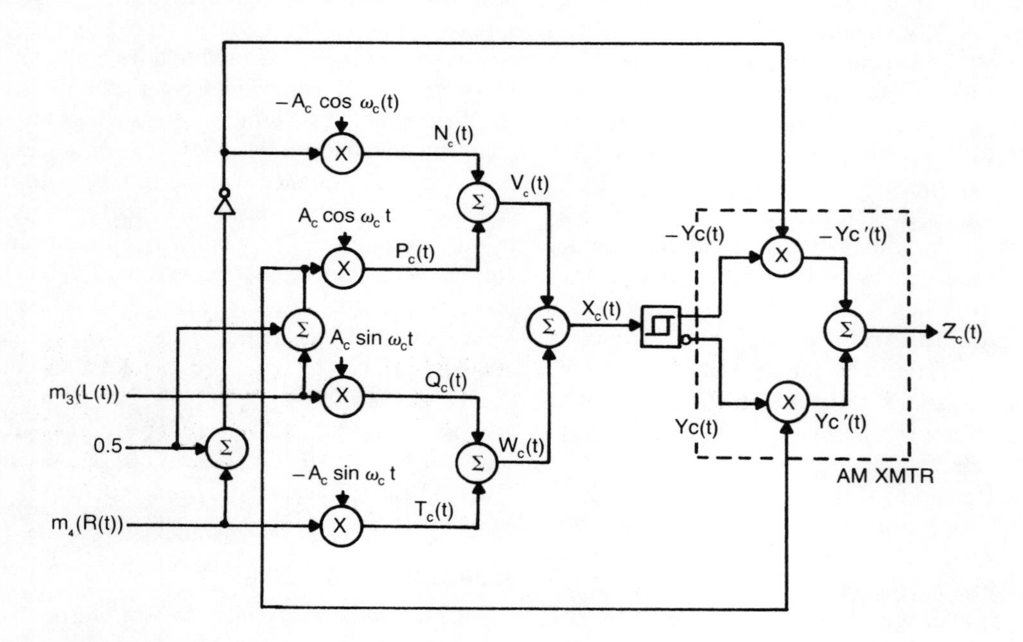

Fig. 4-15. Mathematical model of the AX-10 (courtesy Broadcast Electronics).

Harmonic distortion: at 85 percent mod., L = R; monaural, 0.25 percent max. from 50 Hz-7.5 kHz.

Stereo separation: 50 Hz to 7.5 kHz, 35 dB; 7.5-15 kHz, 25 dB.

Temperature range: 0° to 50° C.

New Modulation Method and Am-10 Operation

For starters, we thought you might like to see how this new Broadcast Electronics modulation works, both in diagram form (Fig. 4-15) and also mathematically. It's pretty good justification for a more advanced system of handling the two channels for better common mode rejection and superior transient response. Symbolically:

M_3 and M_4 are two modulating components.

X_c represents the sum of V_c and W_c and is quadrature modulation.

Y_c represents quadrature phase modulation.

Z_c is C-QUAM modulation.

0.5 at the input amounts to a *constant* for carrier addition and subtraction.

Σ_s are the summations.

X_s are the multipliers.

Note that *both* $A_c COS\omega_c(t)$ and $-A_c COS\omega_c(t)$ reach the transmitter multipliers along with quadrature modulation, producing the C-QUAM modulated output. This is precisely how the two channels are independently controlled for both stereo and monophonic transmission. Left and right channel audio inputs have fully balanced, transformerless amplifiers, and output levels to the transmitter at 600 ohms are level-adjusted between −10 and +26 dBm.

A block diagram (Fig. 4-16) of the entire unit illustrates and extends the mathematical model just shown in practical detail. Beginning at the top, an external or internal oscillator operating at 10 MHz for both frequency agility and as a reference for the 25 Hz pilot tone which may have positive or negative inputs before summing in the digital IF modulator. A mode switching and remote block then accepts ± right and left inputs, which pass through both low pass filters and are 70 percent limited. Signals are then switch-selected for dual day/night transmissions or changing antenna patterns and processed into the matrix/dematrix L and R negative limiter and also to a separate L and R summer and L−R negative limiter for the mono L+R envelope.

Digital matrixing takes place in the IF modulator, which also receives the pilot signal from a low pass filter and a pilot divider. This reference and the quadrature divider are both derived from the 2nd IF divider and its 10 MHz reference master oscillator. When the appropriate divisions reach the frequency synthesizer, the resultant net division is 312 and provides a frequency synthesis of 32051.28 Hz. This is the down-conversion frequency following a preceding 5 MHz up-conversion. The resulting stereo modulated output may then be sampled and/or processed through a limiter and an rf amplifier for rf output. Left, right, L+R or L−R/pilot injection metering is pushbutton selectable through a BNC monitoring jack. At the same time, a LED bar graph continuously displays modulation peaks, and a 125 percent peak-hold detector permits asymetrical modulation monitoring.

STATIONS BROADCASTING AM STEREO

From the best information available a general listing of radio stations broadcasting AM stereo or modifying their equipment to do so is included in Appendix C. We can't attest to total accuracy, but this will be a guide to what stations in the various states and/or cities are offering such service now or plan to do so in the near future. Our sources, of course, are the manufacturers themselves since it would be a monumental task to poll all U.S. broadcasters who might or might not want to say what system they're planning to use if there's severe area competition. We had also planned to publish the pattern area coverages of selected stations, but since AM stereo should encompass the same "footprints," there really isn't any point. In other words, if you can receive normal monophonic broadcasts in certain terrain, the stereo should be available, too, if exciters and modifications have been installed and are producing full frequency and power transmissions.

These listings are also important, not from the standpoint of one or more competing systems, but

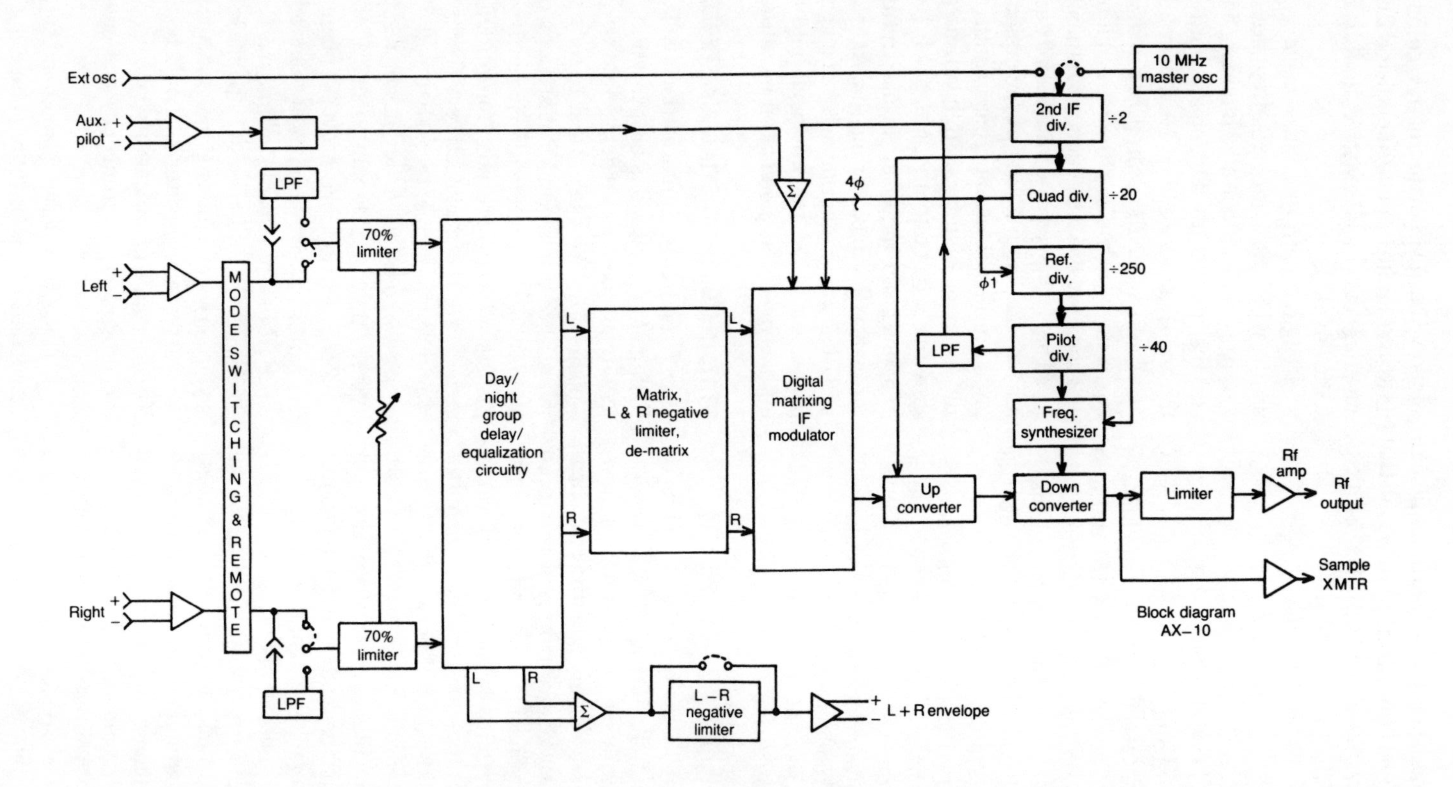

Fig. 4-16. Actual block diagram of the AX-10 exciter (courtesy Broadcast Electronics).

in the future when satisfactory decoders have been developed for the surviving three or four. At that moment, whatever system you're listening to should not be of any concern since you can expect they will all be reasonably adequate.

PROS AND CONS

There are still some unanswered questions about AM stereo such as whose system is best, what are normal or expected system defects, what are likely improvements, and how does it compare with frequency modulation stereo which we've had for so long.

Each of the four currently surviving systems differs somewhat from the others but, except for some high-end frequency curtailment, they are reasonably competitive. Front-runners at the moment are Motorola and Harris, but this could change at any time with either dramatic or worthwhile new developments.

At the moment, co-channel interference becomes a problem when two stations on the same frequency mix signals and produce a low frequency beat note causing modulation and amplitude or phase problems in the receiver. But this interference must be present in significant amounts to be noticed, and one station is normally much weaker than the other because of separating distances. There are also some adjacent (next door) channel effects, especially weak versus strong stations broadcasting in the same vicinity. Some radios have better design selectivity than others—especially the more expensive ones—and these are able to separate stations more easily than the less-expensive ones. Separate-sideband receivers can sometimes tune out such interference, but this can produce intermodulated products sometimes approaching 40 percent which isn't helpful.

At low frequencies, AGC receiver loops may follow the AM portion of any co-channel interference, resulting in intermodulated distortion. You'll notice a rapid up/down volume change, and *all* receivers are subject to this problem. The phase-locked loop portion of any synchronous detector may also follow low frequencies, especially in quadrature modulation involved with co-channel interference, producing more beat notes and fast fluttering. Finally, there's teeter-totter or platform motion that can occur with various systems to a lesser or greater degree. Once again, this is volume flutter when AGC is following AM co-channel and can be heard in *all* AM stereo decoders. If it is a phase-locked loop decoder, there can be an L−R flutter also. This latter, in Harris and Motorola systems, has been called "platform motion" which can also occur in the Magnavox system. In the Kahn-Hazeltine system this flutter becomes a "teeter-totter" effect, slightly different, but appearing in another form. Detuning, unfortunately, causes intermodulation which is an additional problem.

Obviously, all systems have some imperfections, just like television, 2-way radio, and satellite reception. But they all had to start somewhere and there's been steady improvement in each as the years roll by. In the meantime, take advantage of what promises to be another important consumer entertainment plus and enjoy all the advantages of good stereo separation and frequency reproduction without the FM "capture effect" of jumping from weak to strong stations and the added distance AM promises as you drive along the highway. Like CB radio, FM, and 2-way, you're going to have antenna problems and there will be signal cutoff under bridges and tunnels, but then, we know of nothing that's absolutely perfect—at least not in the worlds of mechanics and electronics as we approach the 21st century.

Chapter 5

AM Stereo Receivers

This is by far the most difficult chapter in the book to present in suitable detail because of the numerous and considerable changes that are and will be taking place in both home and mobile receivers reaching the American market. For wideband, satisfactory channel separation, noise rejection, and multisystem stereo decoding, it's very evident that many existing receivers are not now available to do the job. Good designs start with the antenna, continue into the front end, IFs, and finally to suitable detectors capable of recognizing various transmission signatures and efficiently decode them into stereo-audio sound.

All this is especially true in automotive applications where hostile environments of heat, dampness, cold, vibration and shock must be included in the general design, along with workable front end selectivity. Considering cost and utility, U.S. and foreign engineers have their work cut out for them during the next several years if substantially improved receivers are to appear within these constraints. Special emphasis, of course, is on tuners, IF amplifiers and some proposed multi-chip decoders, and even totally new lines of radios. Delco, Chrysler, Concord, Sherwood, Marantz, Jensen, McIntosh, Potomac Instruments, Pioneer, Samsung, and probably Ford have already committed to Motorola's C-QUAM system.

Home radio manufacturers are probably going to be primarily Japanese, with some Tiawanese and Koreans contributing also. At the moment, primary emphasis is on portable (Sony) and mobiles from the three big auto makers and Jensen. Other manufacturers could go for either home or mobile, or possibly both.

INTEGRATED CIRCUITS

Having said all that as background, it's probably wise to begin with some of the integrated circuits currently involved and work up to advanced designs when manufacturers are willing to release details. Radios, too, will be described, at least in block diagram, along with whatever else is currently available, including schematics or other significant technical information.

Motorola's MC13020P C-QUAM Decoder

One of the latest ICs available to detect AM stereo is Motorola Semiconductor's MC13020P 20-pin, dual in-line plastic package which is also the pilot detector. This IC (Fig. 5-1) requires no adjustments, a few external RLC components, and a voltage-controlled oscillator (vco) that operates at eight times the IF frequency with the aid of external resonance. Motorola points out that a crystal-controlled vco is very stable, has a pull-in range of about ±100 Hz at 450 kHz, but can only be used with well-designed front ends. It's pointed out that ceramic and LC vcos have pull-in ranges of ±2.5 kHz at 450 kHz, and ceramic devices accurate enough to avoid trimming are available with matching capacitors. The RC network in parallel with the "crystal" becomes a phase-locked loop filter and sets the loop corner frequency of 8-10 Hz, in lock.

This Motorola IC (Fig. 5-2) receives its input through a coupling capacitor that supplies an intermediate frequency of some 455 kHz to the variable gain block as well as to the envelope detector. L+R is immediately developed as the monophonic signal, while the output of the envelope detector and the I detector are compared by the error amplifier to reduce C-QUAM to QUAM. This, of course, is quadrature, and the I and Q detectors are secured at 0 and 90-degree demodulation angles by references from the phase-locked loop, divide-by eight vco.

The level detector now senses carrier level and can be used for an AGC source as it controls gain of the L−R signal so that IF level changes of ±6 dB can be controlled within ±1 dB. This means that the pilot tone detector will receive the 25 Hz signal with little change in amplitude. In a receiver, Motorola recommends a 2.5 percent threshold level for pilot tone detection, so that as much as a 2:1

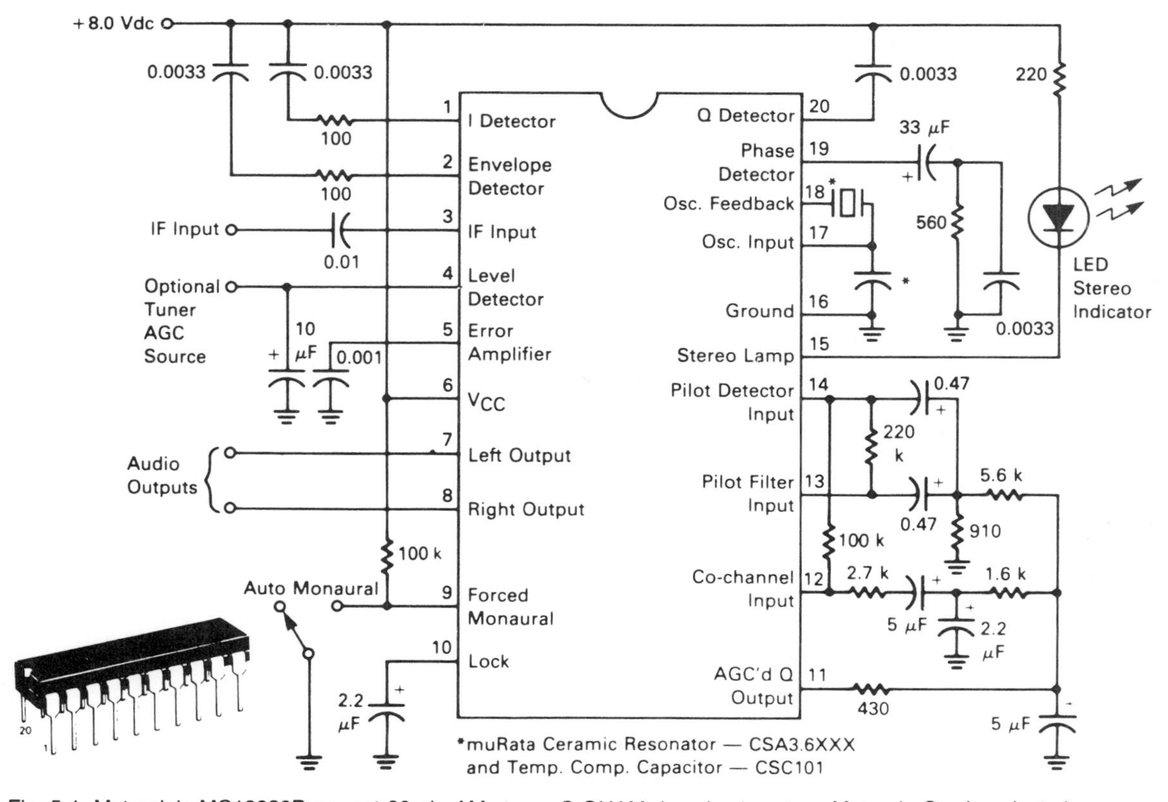

Fig. 5-1. Motorola's MC13020P newest 20-pin AM stereo C-QUAM decoder (courtesy Motorola Semiconductor).

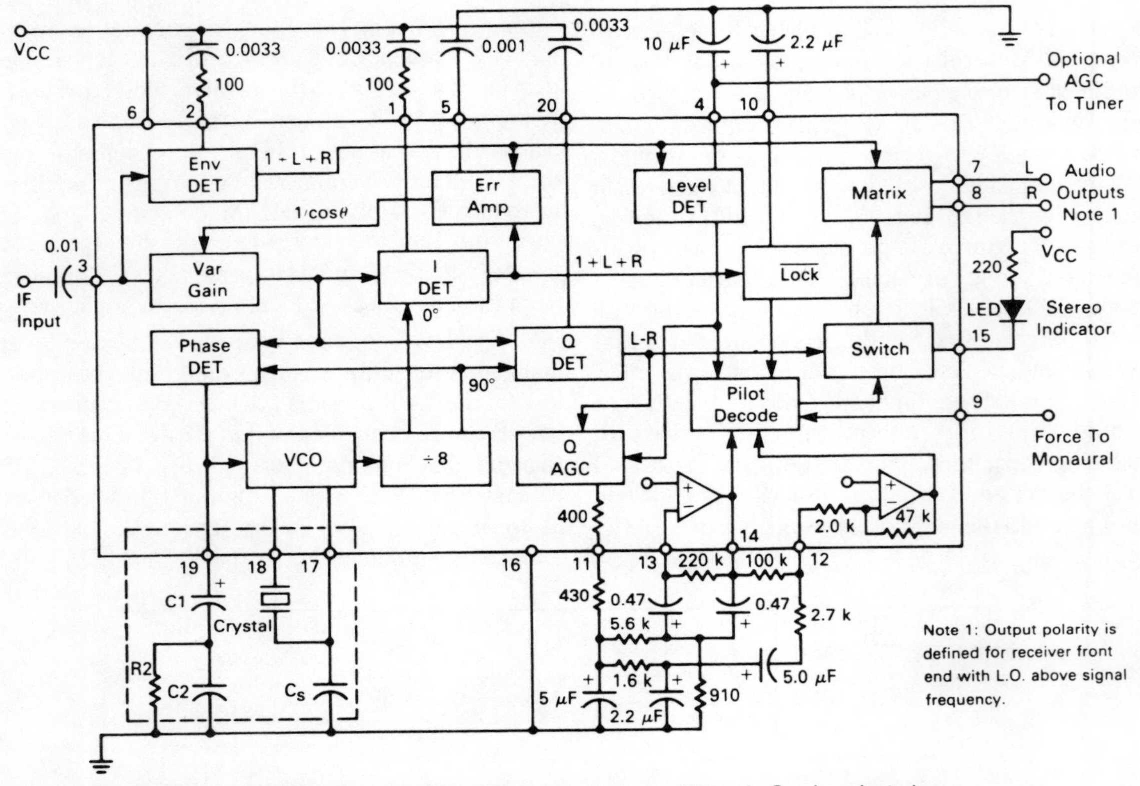

Fig. 5-2. A functional block diagram of the C-QUAM decoder (courtesy Motorola Semiconductor).

reduction in detected pilot tone level due to external component tolerances will still permit adequate decoder stereo mode operation. The lower cochannel input absorbs any low frequency intercarrier beats and prevents the pilot decoder from detecting stereo at certain levels.

Motorola's pilot decoder has two modes : upon satisfactory reception the decoder will turn on stereo following seven continuous 25 Hz pilot tone cycles; but under poor signal conditions, the pilot counter must count 37 cycles before permitting stereo. Motorola says that each such disturbance will reset the pilot decoder counter to zero. The level detector will also keep the decoder from stereo operation if the IF level drops 10 dB, but won't affect the pilot counter. In the stereo mode, the co-channel input is disabled and, therefore, the decoder cannot automatically switch out of stereo because of co-channel action. Nonetheless, the I detector will cause the stereo decoder to revert back to monophonic at about 14 dB S/N with respect to 100 percent modulation. During stereo, any apparent noise is detected by negative excursions of the I detector. Then, when all pilot decoder inputs are reasonable, the switch is turned on, delivering L−R to the Matrix and the entire MC13020P operates, delivering left and right stereophonic material to final amplifiers and speakers.

Multiplier-Detector. What those of us who are videophiles used to calling synchronous chroma detectors in television receivers has a long history—even dating before 1970—and other important uses as well. This is especially true for Motorola where its linear amplitude detector is actually termed a doubly balanced multiplier (Fig.

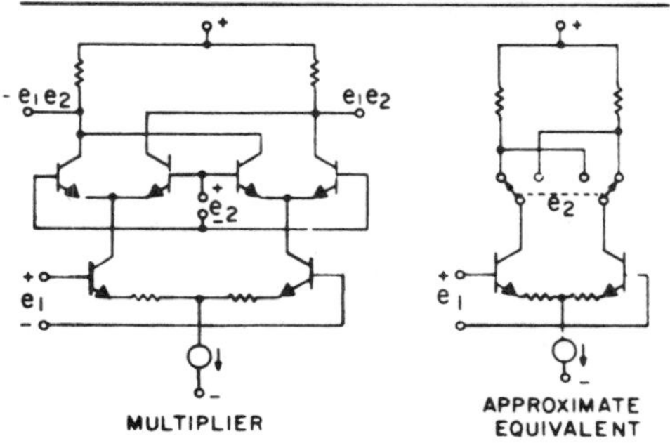

Fig. 5-3. Motorola's current source-balanced multiplier (courtesy Motorola).

5-3), consisting of positive and negative inputs into current source transistors appearing as a differential amplifier and serving a pair of synchronized, single pole, double throw switches. Switching occurs because of e_2 and the outputs are full wave multiples of inputs e_1 and e_2.

Such inputs are out-of-phase current generators supplying excitation to the "ideal" switches, producing an output equal to the product of the input(s) and the switch drive, whose zero crossings occur because of the zero crossings of e_2. After filtering, output voltages contain only the input voltage product and amplitude levels from the ± switch.

As you can see in Fig. 5-4, multipliers may readily be used for FM detectors, square law detectors, or linear detectors. Of course here we are concerned with the multiplier linear detector and a version of this circuit appears in Fig. 5-5. As Motorola explains, in this configuration you may also add external emitter degeneration to the input differential amplifier and also adjust current drain as desired. The diagram shows all prior inputs, two additional transistors supply stages, a transistor acting as a diode, resistor degeneration sources and probably a dc variable source at terminal 5 for accurate biasing.

National Semiconductor's LM1981

This is the original IC designed for Magnavox's AM/PM system which has L+R envelope modulation and linear phase modulation for L−R. In the meantime, National has adapted this same design to other emerging systems and has even offered information on a "universal" stereo receiver that would have sufficient bandwidth, low noise, and rf AGC. First, however, let's describe the LM1981 and then we'll attempt to put the LM1893 and LM1981 together in one unified package which, according to National, can process and decode Magnavox, Harris, and Motorola systems.

A dual in-line 20-pin package for the LM1981 AM stereo decoder is illustrated in Fig. 5-6. You can readily see the IF carrier input, filters, holding capacitors, limiter and PM detector input, pilot tone output, and the rest. Rf frequencies between .55 and 1.6 MHz are heterodyned down in a superhet front end to either 262 or 455 kHz IF frequencies. Here the IF is set at 455 kHz and 400 mV when rf signals are above AGC threshold. But the input can handle signals +6 dB greater or −20 dB less than 200 mV for adequate stereo reproduction.

Prime features of the LM1981 are illustrated in Fig. 5-7. This IC is designed to decode stereo that's both amplitude and angle-modulated on an AM broadcast carrier. The IF input is delivered through an input amplifier to the L+R AM detector and also through a limiter amplifier to the adjustable PM detector and its regulator for eventual L−R resolution. The AM detector is full wave and con-

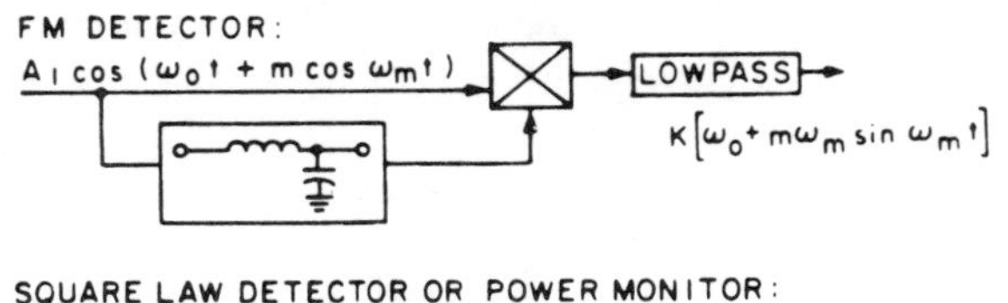

Fig. 5-4. Various Motorola multiplier Configurations (courtesy Motorola).

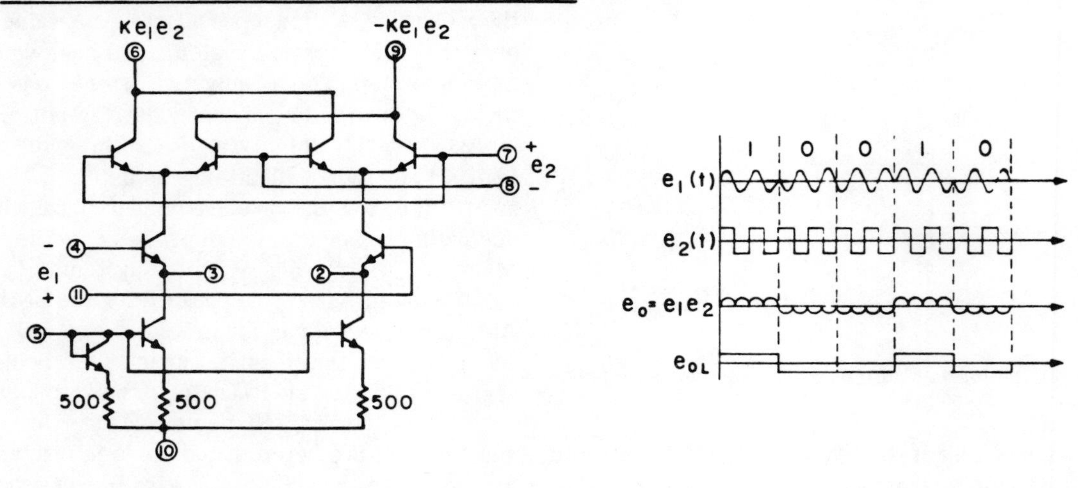

Fig. 5-5. A monolithic IC example of a multiplier and active waveforms (courtesy Motorola).

sists of two transistors in differential configuration with active loads, dual inputs and outputs. In the PM (angle modulation) detector, signals are subject to five stages of (AM) limiting before reaching the detector. The rf limited carrier then switches one set of balanced inputs to the multiplier while others are receiving signals after they are routed through a resonant circuit tuned to carrier frequency.

Now both multiplier input ports are switched, and conversion gain is controlled by the parallel variable resistor across the multiplier which, in schematic form, appears very similar to a color television chroma detector with signals entering two pairs of current drivers feeding dual pairs of switched and phase-shifted detectors above. These switch collectors, however, are RLC tuned, whereas Chroma demodulators are matrixed.

Connected across the multiplier output is an active inductor consisting of two operational transconductance amplifiers and a low pass filter. This places a dc "short" across the multiplier outputs preventing offset voltages from exceeding the "dynamic" range of the following stage(s), and also produces a low frequency pole whose frequency is about 30 Hz. With the capacitor value chosen at 0.047 μF, the inductor would have to be almost 600 henries, and so an active inductor is well suited to the application.

Detected L−R goes to the matrix through a variable gain block that depends on the average level of the IF carrier. It then passes to a second gain block controlled by the mute/blend input permitting stereo/mono switching or stereo/mono blending when rf signals are insufficient. L+R and L−R are now matrixed and outputs go to dual sample and hold circuits for right and left stereo outputs.

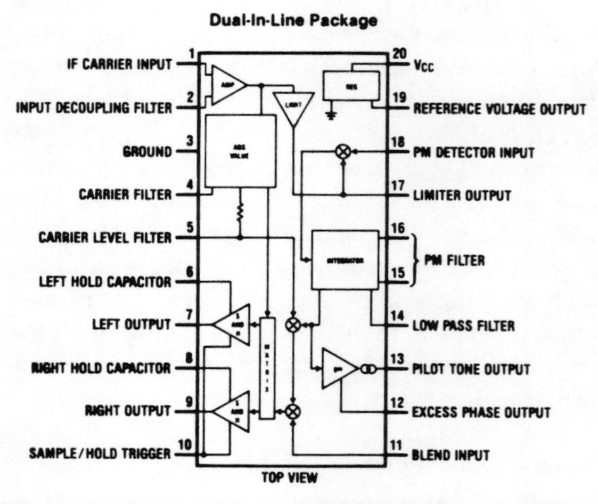

Fig. 5-6. National's 20-pin LM1981 contribution to AM stereo detection (courtesy National Semiconductor).

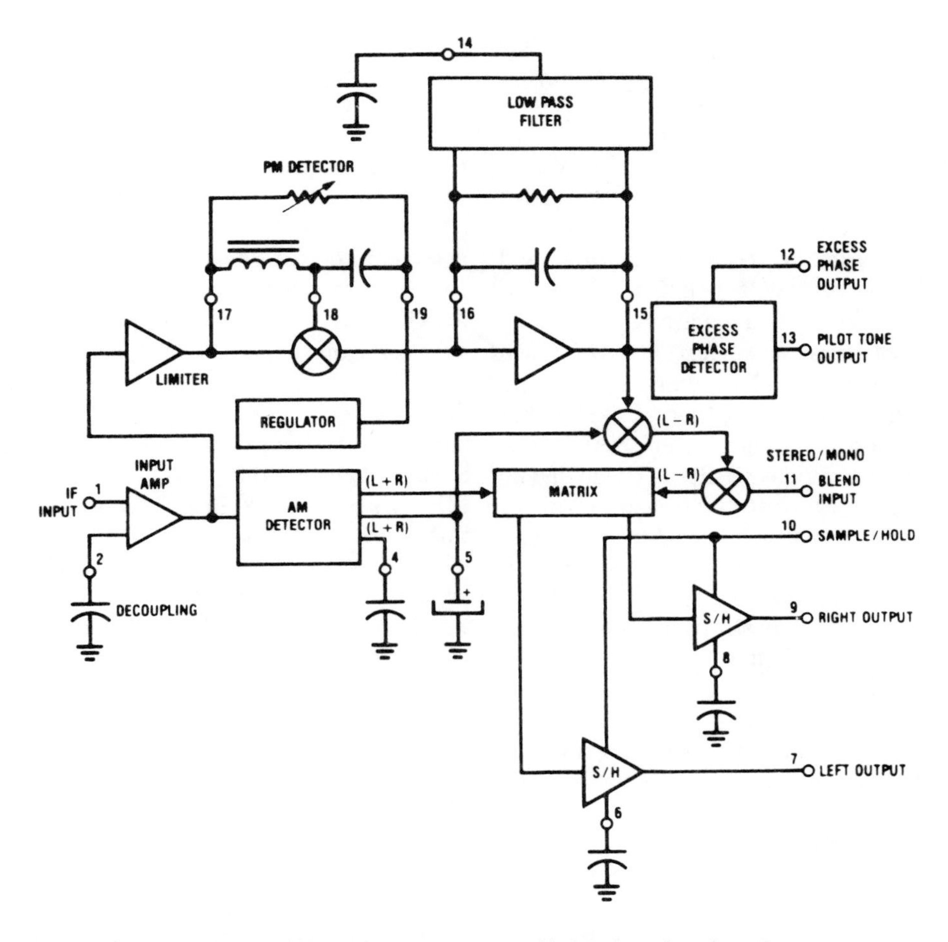

Fig. 5-7. Overall block diagram of the LM1981 decoder (courtesy National semiconductor).

The excess phase detector picks up signal current from the L−R before the two gain blocks and as peak phase deviation increases, produces differential current to turn on the stereo tone detector.

That pretty well winds up a fairly intensive description of National's LM1981 stereo decoder, which we'll probably see more of as refinements and receiver manufacturing begin in earnest during 1985.

National's Complete AM Stereo Receiver

Rather than try to modify some of the standard AM receivers still on the market, National's Fred Cheng and consumer linear applications manager Martin Giles have tied together an LM1893 "front end" to the LM1981 decoder for what they call "a universal AM stereo receiver."

In so doing, they investigated whip and ferrite rod antennas, finding that the former will supply as much as three times the available signal to an AM receiver's tuned rf stage. But they also found that in order to reject undesirable image frequencies while passing an acceptable signal, a double-tuned rf stage was required, and then an additional rf amplifier added to make up for the tuned circuit loss. The mixer, they discovered, needed minimal redesign since most contemporary IC mixers are double-balanced and usually reject stray IF formed because of reduced rf Q. Delayed AGC for the rf keeps mixer overload and cross modulation under control. But if phase instability arises in the local oscillator driving the mixer, then stereo decoder detection sees this as noise, distortion, or loss of stereo separation. Therefore, well designed oscillators limit such problems to some −60 dB. Varactor diodes, they say, cause a 3-4 dB S/N loss; receivers using mechanically variable tuning capacitors are subject to microphonics; and maximum S/N ratios range between 45 and 50 dB for quartz crystal controlled electronic tuners.

IF selectivity was tackled next and they found that tuned LC or ceramic ladder filter circuits "typically have −6 dB bandwidths of only ±3 kHz." But this noise and adjacent channel interference tactic doesn't help full bandwidth receivers occupying 30 kHz. But, then, only very strong transmitters (say 50 kW) would deliver a usable signal to such wideband receivers. While 10 kHz channel spacing offers adjacent channel protection during the daytime, and along the ground wave, skywaves at night multiply the number of interfering stations and produce the familiar "garbling" anyone can hear by just attempting to tune in local transmissions which are often overridden by distant ones bouncing from the ionosphere at 80 kilometers above the earth. Then there's the space wave, representing energy between the transmitting and receiving antennas within the earth's troposphere, 10 miles or less above the earth's surface. So the time of day is a factor in AM reception that must be taken into account.

Cheng and Giles then state that practical AM radio measurements indicate ±8 kHz for daytime and ±4 kHz for nighttime IF filter bandwidths, and the need to develop wider bandwidths for ± 10 kHz response rejects cheap IF transformers "because of coil Q limitations." They foresee coming AM IF filters as ceramic-based devices, having inductors only to suppress undesirable sidelobes. But there's also passband ripple, sideband amplitude symmetry, and phase response to plague stereo reception. Therefore, they now suggest a totally different rf and IF approach to the problem by inserting a recently developed IC into the receiver that, according to them, cures most if not all the evident ills.

The LM1893, it is said, does the trick! Figure 5-8 tells the story of what this particular IC is all about. Note the relatively few coils, only an occasional diode and the usual complement of outboard resistors and capacitors, with an AGC-controlled FET front end. Certain manufacturers may or may not use this precise design since it's primarily a demonstrator but, usually, if two ICs make a complex radio then a good deal of attention will be paid to what they do since chassis size and cost are always major factors.

After the antenna and FET, the front end is double-tuned and then followed by a bipolar amplifier supplying drive for the mixer input. Such double tuning produces 28 kHz bandwidth at 1.2 MHz, with an 80 dB image rejection figure. Part of this same signal also reaches the AGC detector, which then processes levels for rf AGC control, preventing overloading and mixer crossmodulation. You also see a local oscillator output whose phase noise, we're told, measures better than −60 dB at 1 radian deviation. Then there are the LC tank circuits, tuning meter, a 450 kHz resonator, mixer output, 450 kHz filter, and IF input-output that eventually winds up at the previously discussed stereo decoder. Station tuning, of course, is by voltage-controlled varactor diodes.

When you put all this together for the three systems, Magnavox, Motorola, and Harris may all be decoded (says National), by the arrangement depicted in Fig. 5-9. Here you see a few resistors and capacitors added after the L−R detector and in parallel with the integrator, plus a variable potentiometer and shunt capacitor inserted after the full wave L+R detector for Harris and C-QUAM. The pot following the FM limiter permits L−R and L+R level matching to secure maximum stereo separation. The integrator, of course, produces a phase modulation signal for the two multipliers and the L and R output matrix. You will, however, need a separate pilot tone detector for stereo lamp indication. Mssrs. Cheng and Giles complete their explanation by stating that "single chip IC decoders to date can detect pilot tones for specific systems, but good multiple pilot tone detection schemes have yet to appear." Stereo indicators, of course, are

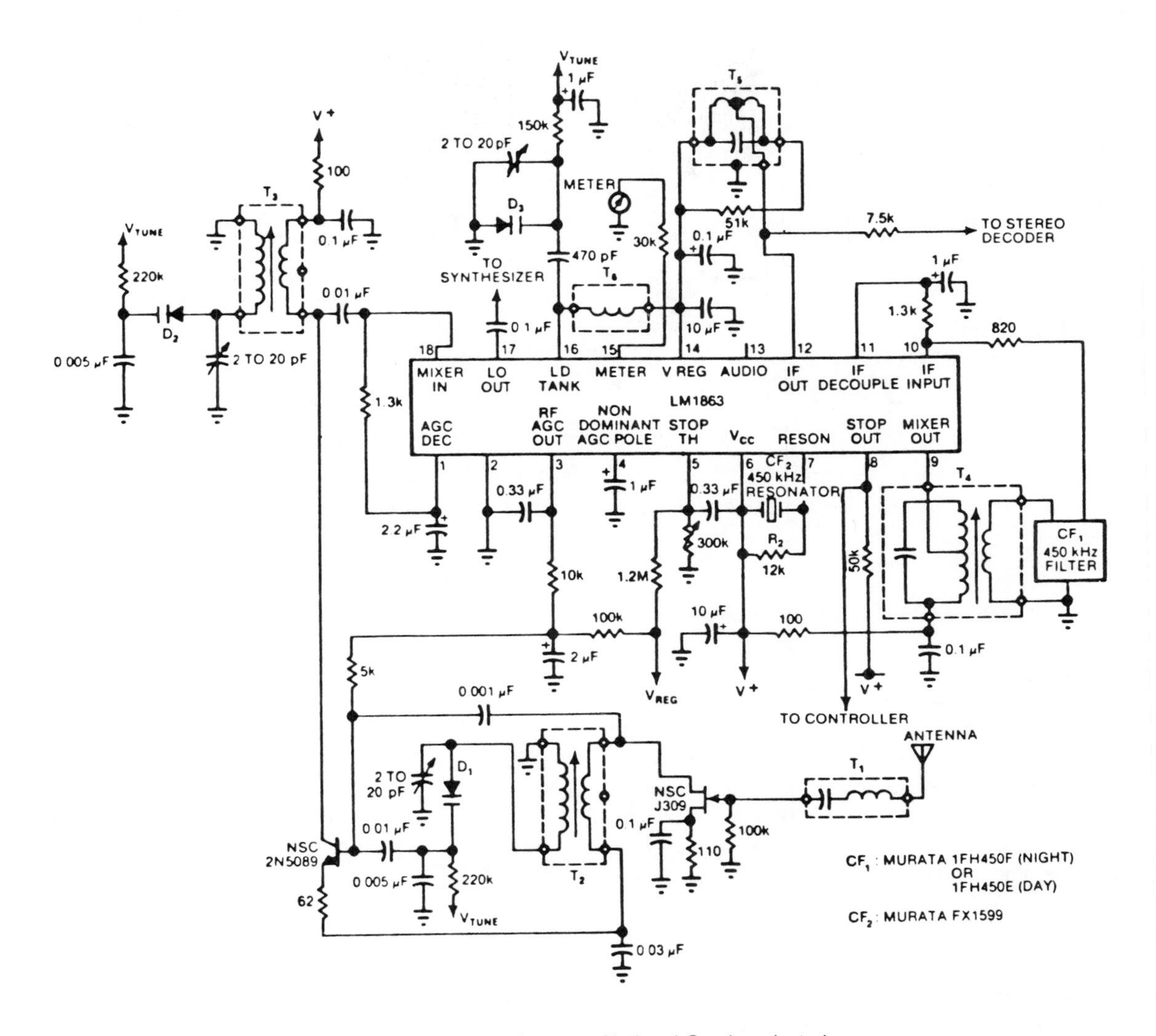

Fig. 5-8. The LM1893 front end wide band receptor (courtesy National Semiconductor).

especially important for motorists on the move who normally require a stereo identifying signal.

Although this particular diagram does not show that Kahn/Hazeltine's system can be decoded by the same general principles, we're told that the Magnavox MPX is very similar except that detected L+R and L−R have been phase shifted by ±45 degrees. So what's needed here is a pair of 90-degree phase shifters following the matrix so that the 45-degree angles may be compensated and left and right signals recovered. Of course, we've not seen one particular IC that does everything, but one may well appear.

Motorola, by the way, disputes the claim that the LM-1981 decodes C-QUAM correctly and believes that current and projected 4-way decoders compromise performance in *any* of the systems. A single system, they say, can be more easily designed and implemented for best results—a premise they seem to be winning.

Electronic Tuners

While we've discussed signal processing in

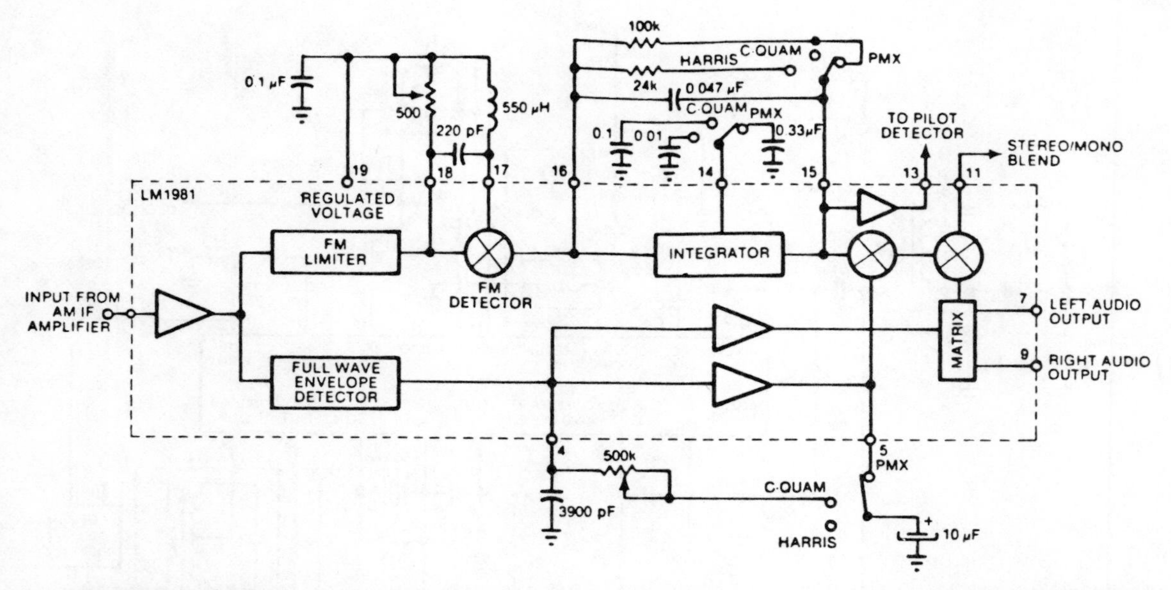

Fig. 5-9. Mating the LM1893 and LM1981 produces a receiver to decode major AM stereo systems (courtesy National Semiconductor).

AM stereo radio front ends, little has been said thus far about tuner control . . . and by this, of course, we mean digital tuners. On some of the more expensive AM/FM radios, we already have pushbutton signal-seeking station selection, liquid crystal time and station frequency readouts, and memory for more than a dozen AM or FM stations which may be programmed in or out of memory at will. The next step, obviously, is automatic signal-seeking, and this, too, is now available in deluxe models. Its use will spread with reductions in cost. All this naturally, is very important to automotive mobiles for both safety and driver convenience, and may well spread to component radios as home communications centers grow in both utility and importance. Perhaps one day there'll be modular plug-ins for cellular, AM/FM stereo, cable, satellite, TV, and VCR/disc, all in one, unobtrusive console. Electronically, that's possible right now. Practically, the market may not be quite ready for the cost of such somewhat complex equipment.

As the use of ingenious, multi-function, and repeatable/reliable integrated circuits grows there will be many changes, especially where analog has been converted to digital, with all its advantages of less noise, signal reliability, frequency separation, secondary image cancellation, simpler troubleshooting, and eventual reduction in cost. At any rate, Donald Wile and National's manager of consumer applications, Martin Giles, have put together some interesting thoughts on new ICs for electronic auto radio tuning that offer worthwhile consideration. These we'll paraphase with additions, while using their excellent diagrams as illustrations.

A good place to start in any keyed or signal-seeking tuner is the front end, and especially the mixer. Here incoming AM or FM signals are amplified to some usable level and then routed to a mixer where they are heterodyned down to an AM intermediate frequency of 262/455 kHz, or 10.7 MHz for FM. This then, permits all incoming frequencies to be reduced to a constant intermediate "carrier" that still retains all the original modulation but remains fixed for further amplification and baseband extraction.

The key to received frequency stability and positive tuning is the local oscillator (LO) and mixer combination where incoming radio signals for AM between .55 and 1.6 MHz and 88 to 108 MHz for FM are initially detected and reduced to their inter-

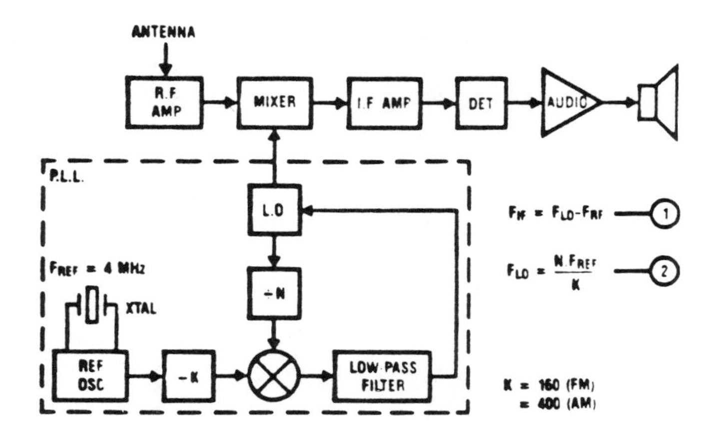

Fig. 5-10. Phase-locked loop portion of digital tuning (courtesy National Semiconductor).

mediate frequencies (below rf but above audio). For accurate station selection and non-drift tuning, this is now done by phase-locked loop circuits consisting of a local oscillator, divide by N counter or reference, a comparator, a crystal-controlled reference and another down divider. As channels are tuned or possible drifting occurs, local oscillator and crystal oscillator outputs are compared and any difference is filtered and returned to the local oscillator as a dc correction voltage, forcing the local oscillator to track correctly (Fig. 5-10). This is called synthesized tuning (from frequency synthesis), and a control IC may now be used to set the PLL dividers and deliver serial digital drive to a visual frequency display. The operator may then engage pushbuttons for selected radio stations or allow signal processors to "scan" incoming signals and deliver a *halt* command when a strong enough rf carrier is detected. Usually 50 milliseconds is enough for station identification and then an 8-second pause takes place on identified stations before scanning continues or until some particular station is chosen and scanning stops.

All this is illustrated in Fig. 5-11 where you can see such blocks as Valid Station Stop Detector, AM/FM/PLL Frequency Synthesizer, Keyboard, IC Controller, and Display. The Controller times all stops on the basis of signals from the Valid Stop

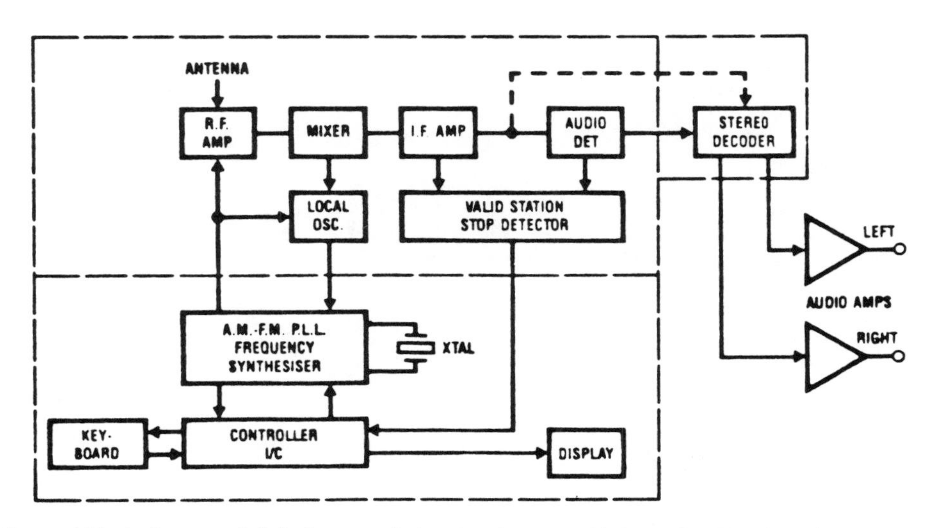

Fig. 5-11. General block diagram of digitally-controlled tuning (courtesy National Semiconductor).

Detector which has inputs from both the IF amplifiers and the audio detector. Any station selected, of course, has to be signalled by the keyboard either from memory or on the basis of adequate rf.

For FM, the tuner gain measures between 30 and 45 dB at a 5-6 dB noise figure (Fig. 5-12), and while the rf stage offers some image frequency and adjacent channel interference rejection, greatest selectivity comes from a filter block between tuner and buffer. Filter impedances and the IF gain block may be matched to reduce both insertion loss and any passband ripple.

The buffer amplifier permits the IF increased gain as well as front end IF isolation and better signal sensitivities. Capacitance, however, must be carefully dispersed to prevent any unnecessary feedback in all these high-gain circuits and any coils should be physically separated from all other portions of the system where there are possibilities of undesirable oscillation.

As for the AM tuner, designers are finding that not only does the detector and IF-AGC stages require considerable rework, but also the tuner itself, including the addition of a field effect (FET) transistor front end (Fig. 5-13). Here you can see the obvious capacitance tuning of T1 and T2, along with the secondary inductance of T1 and the primary inductance of T2. As National notes, the usual signal path from FET Q1 to bipolar amplifier Q2 is through capacitor C19. But for AM, C26 isn't a total bypass to ground, and a certain amount of signal also reaches Q2 though capacitor C18 when the radio is tuned to the higher frequencies.

On low band, Q1-Q2 gain is reduced by T1-T2 reflected impedances and more signal passes through C18, compensating for this gain reduction, reducing gain change between .5 and 1.6 MHz to less than 6 dB instead of the usual 14 dB, which demonstrates very clearly why mobile radios require the most careful design for maximum range and broadband stereo efficiency. Simply modifying some of the better AM monophonic radios in the beginning will put AM stereo on the road, but better

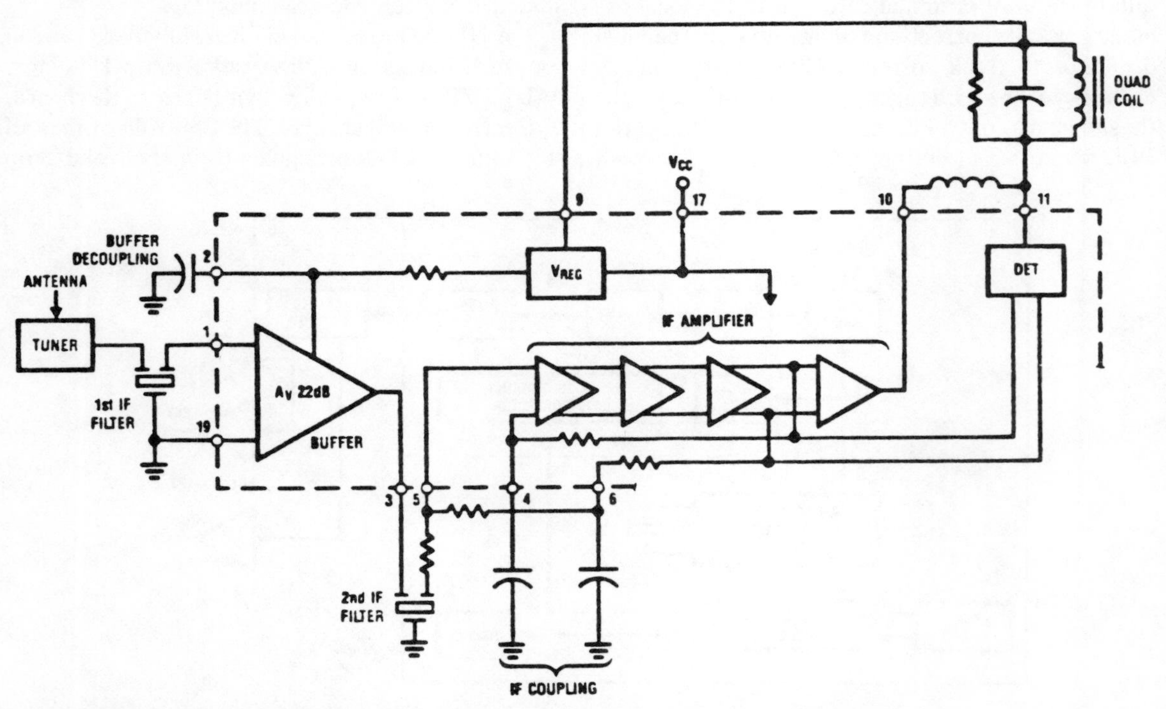

Fig. 5-12. Radio selectivity depends on filter block and IFs (courtesy National Semiconductor).

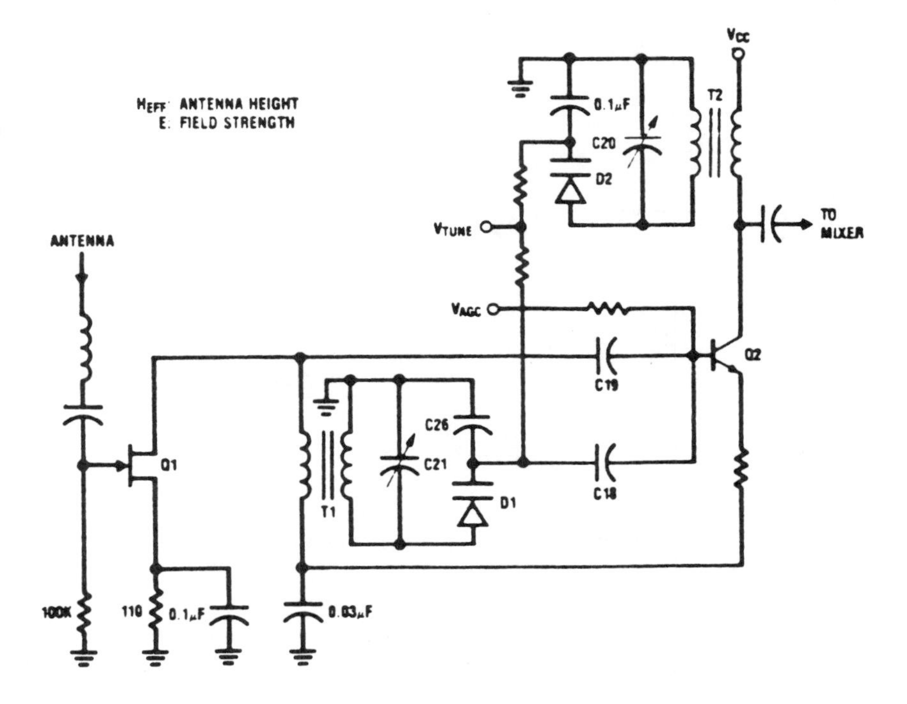

Fig. 5-13. New tuner circuits needed to process AM stereo (courtesy National Semiconductor).

results should be forthcoming as engineers dig further into the problems and come up with satisfactory solutions. We do expect competition to make a great deal of this possible, and fairly quickly with the help of more and better integrated circuits.

The Stereo Pilot Signal

We're speaking about the AM stereo pilot signal rather than the much faster FM version, which chugs along at 19 kHz. Here these frequencies are below 100 Hertz, and in the case of Magnavox, 5 Hz. So pilot tone detection in some receivers certainly requires between 280 milliseconds and seconds, not counting the phase-locked loop settling time of as much as 500 milliseconds. Once again, Fred Cheng of National Semiconductor has done considerable work in this sector, and here's a partial report of what he's found.

Some AM receivers can generate noise of considerable magnitude in the range of pilot frequencies that could possibly switch mono to stereo, or the reverse, flicker the pilot lamp, or false-trigger a stereo detection circuit. (On the other hand, Motorola claims that any low frequency noise in its receivers is generated by poorly operating synthesizers, but such noise rarely affects the MC13020's pilot tone detector.) You also want to at least identify some AM stereo station even when the signal is less than threshold. So let's see how these problems were overcome in the Magnavox system (Fig. 5-14).

Here the carrier is phase modulated with a 5 Hz level of 4 radians and current of 28 microamperes, an impedance of 48K ohms, and a resulting pilot level of 1.3 volts peak-to-peak. To deliver a relatively noise-free, high quality pilot detector, the incoming 5 Hz signal is initially processed through a high gain and low Q bandpass filter centered at 5 Hz. This both improves S/N but also limits and squares the pilot to prevent any false triggering from noise. This improves pilot detector sensitivity, permitting 5 Hz signal recognition even though it is actually *below* AM stereo threshold.

As stations are tuned, it's possible to place a 5

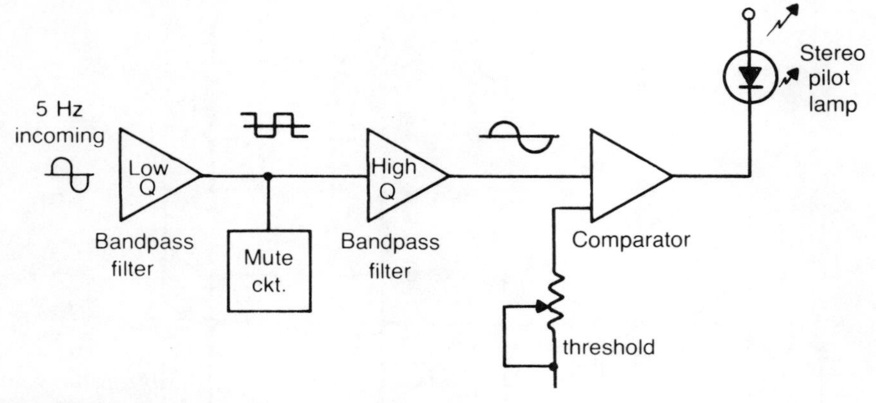

Fig. 5-14. Pilot detector block diagram.

Hz (or so) signal on the PM detector output and trigger a lighting of the stereo indicator. Therefore, a muting circuit must operate to prevent false triggering, and this is done by a transistor switch placed across the detector. High voltage pulses, of course, turn the transistor on, shunting them out of the circuit.

Next, a high 10-20 Q bandpass filter, also centered at 5 Hz, removes all low frequency noise by its sharp response. The resulting signal then proceeds into a comparator, and when it reaches this circuit's threshold the stereo indicator lamp is lighted. Pilot detection time is mainly determined by the Q of the bandpass filter and amounts to approximately 1 second.

Hazeltine's Multiple Decoder

Stating that "while broadcasters can make their system decisions on a technical basis, based on performance in the broadcast environment, receiver manufacturers are largely at the mercy of market conditions and the direction the broadcasters take", Hazeltine has announced a "building block" IC chip for all four surviving AM stereo systems. The company believes such a decoder would make it easier and safer for receiver makers to enter the AM stereo marketplace, instead of risking everything on a single system.

An overall block diagram of the integrated circuit—to be tested and evaluated during the summer and fall of 1984 has been released by Hazeltine, but without any technical description. It appears in Fig. 5-15. What we see on paper are received signals reaching an IF signal processor, with feedback from the monophonic L+R detector and a link between pilot detector and processor meaning, undoubtedly, pilot signal demodulation common to all systems. The stereo L−R detector is traditionally regulated by a phase-locked loop source, and both L+R and L−R information travel to audio processor I and, sequentially, processor II and III. A secondary PLL signal also seems to operate on audio processor II which receives pilot signals from the pilot detector and processor. Outputs are the usual left and right stereo-demodulated and stereo pilot indicator signals.

There's another block diagram, not shown, that seems to accompany the system block describing both a simultaneous detection and evaluation and a time sequential detection and evaluation method. Time sequential uses a tunable filter before a single L−R detector, followed by four evaluators and a six-way logic output that includes mono, stereo, and A through D, probably representing the four decoded systems. The simultaneous detector shows L−R entering four filters, followed by four detectors and four evaluators, and following the same type of logic strip with equivalent outputs to time sequential.

Our crystal ball divines the L−R breakdowns as two methods of achieving the same stereo goals.

And although the simultaneous method appears to have a higher parts count, the tunable filter and its control in time sequential may well mean considerably more complexity. Undoubtedly we'll know a great deal more once Hazeltine has proved the design and the unit has become fully marketed.

EXISTING RECEIVER MODIFICATIONS

We'd like to include whatever information on the subject becomes available before we go to press, so that those who might profit from its inclusion will have a reasonable idea of the various procedures. This is decidedly *not* a suggestion that everyone begin modifying all sorts of AM radios so they'll hear good-sounding stereo. As you have already surmised, only a relatively few good or better equipments will permit modifications for all sorts of reasons, and the engineering required probably stumps all but the audio professionals who have access to a great deal of software and hardware design tools, and a generous bin of appropriate parts. But there's no law against trying, if the spirit so moves and time presents nothing in the way of an insurmountable obstacle.

Actually you may want to sharpen your latent skills to actually see if you're as good as you think you are. Most of us have always found that anything worthwhile in life presents an interesting challenge and almost never comes easily. So have at it if both the spirit and pocketbook are resolute! Naturally, we have to work with those supplying us material, so we'll start with Motorla's C-QUAM decoder MC13020P and proceed from there. A Technics SA222 is the vehicle.

Technics SA222 For AM Stereo

Motorola's assumption is that this receiver has enough genuinely solid design to accept AM stereo and reproduce it with reasonable volume and fidelity. A partial schematic appears in Fig. 5-16. Of course, the FM section is also an integral part of the whole, appearing above the AM section until both AM and FM are joined in IC101 where their respective intermediate frequency amplifiers begin the process of enlarging both signals for eventual audio detection. To avoid adding all sorts of special output amplifiers, the FM section, of course, must be stereo so that two channels of sound are readily available. But remember that these final audio stages only have to handle a bandspread of between 30 and 15,000 Hz. So the difficulties of L and R detection, channel separation, and initial amplifica-

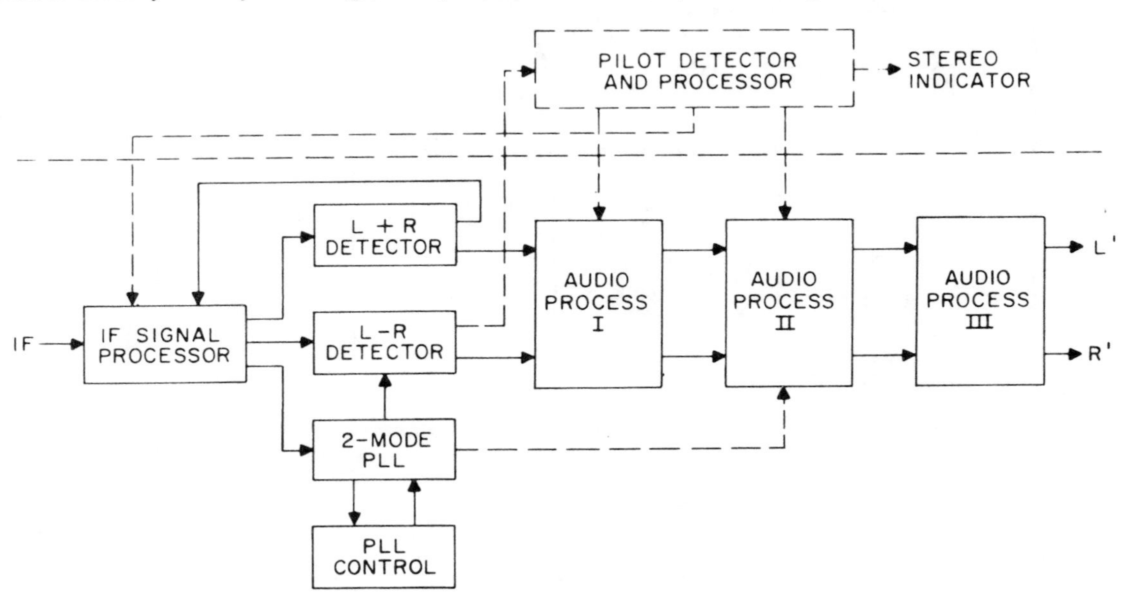

Fig. 5-15. Functional diagram AM stereo multi-system compatible building block IC (courtesy Hazeltine, Inc.).

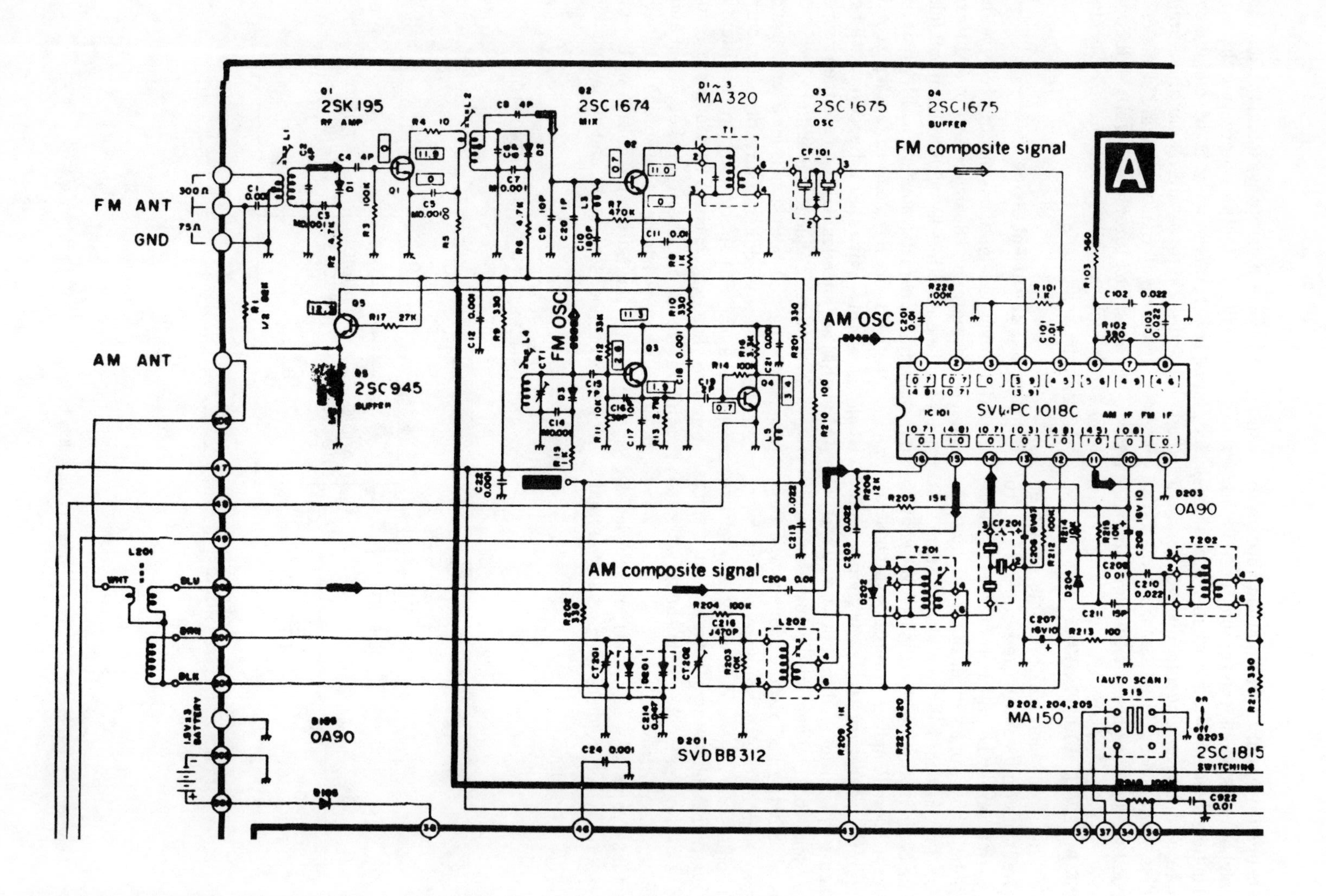

Fig. 5-16. Technics SA222 receiver before modifications (courtesy Technics).

tion have been overcome before speakers are eventually driven by these power outputs.

The immediate problem in such a modification, then, is the AM detector (Fig. 5-17). The procedure begins by wiping out π filter Z201, changing bias around D203, then adding a pair of complementary NPN-PNP transistors in dual emitter configuration of Q202, Q203, etc. Because the primary voltage of T202 causes transformer IPM, its level must be reduced, and the new circuitry added. Next the synthesizer is modified to decrease low frequency noise by changing the value of loop filter C902. The 1 megohm resistor across C901 is added to create leakage current in the loop filter to lower synthesizer "dead zone" noise.

After cutting the circuit board foil and adding the transistors and their bias circuits, the stereo decoder can be installed. IF filters are important and Motorla recommends a Toko 450 CFMS unit with a ±500 Hz center frequency tolerance option. You'll also need an 8V regulator for the IC and, apparently the same three transistors since their loads are all referenced to the same operating potential. The rest of the basic circuitry is shown in Fig. 5-18. For best stereo, Motorola recommends 10 kHz notch (whistle) filters connected to left and right stereo outputs. The transistor connected to Pin 9 is used in time delay for fast pilot detection.

Parts lists changes are as follows:

☐ *Remove from Turner PC Board* Z201, D203, CF201, C217, R216, R217.

☐ *From FM/AM Preset Tuning Circuit PC Board* remove C902 and replace with 3.3 μF electrolytic and add 1 megohm ¼W resistor across C901.

☐ *Miscellaneous*: Add 12-ohm resistor between brown lead and pin 201 off loop stick top, and two 2.2K resistors between T201 and CF201.

And when all's securely tacked aboard, you should hear the "sweetest music this side of Heaven."

Pioneer SX-6

If the foregoing is electronic music, let's take another Motorola conversion—this time Pioneer's SX-6 and see what can be done here. Unfortunately, schematics for this one are just bluelines in large folders and we'll have to reduce them to book dimensions as best we can. Nonetheless, you good engineers will get the idea and be able to fill another circuit board with extra goodies and come up a winner, if you so choose. This is a high quality, high fidelity, in-home stereo receiver capable of 45 W output at only 0.009 percent THD distortion. The AM tuner, however, has narrow bandwidth because

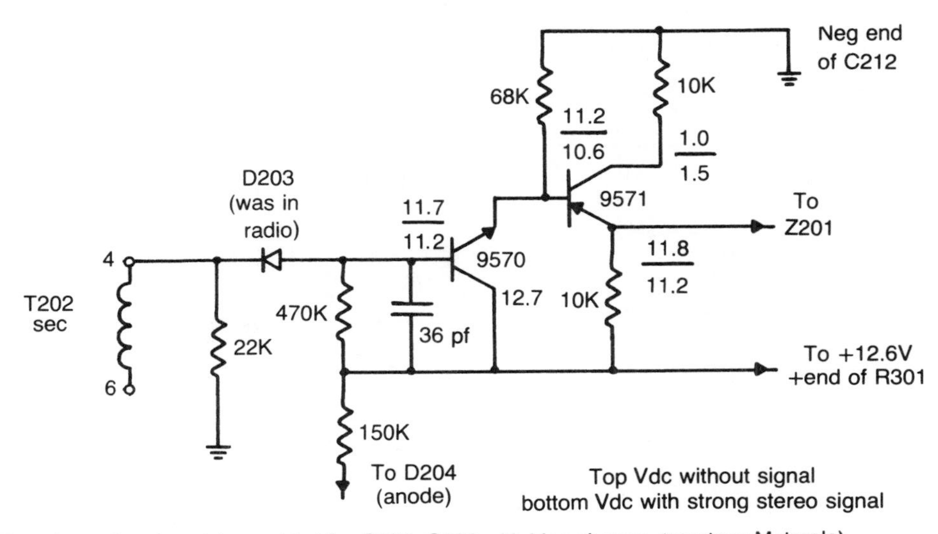

Fig. 5-17. Complementary transistors added for Q202, Q203 with bias change (courtesy Motorola).

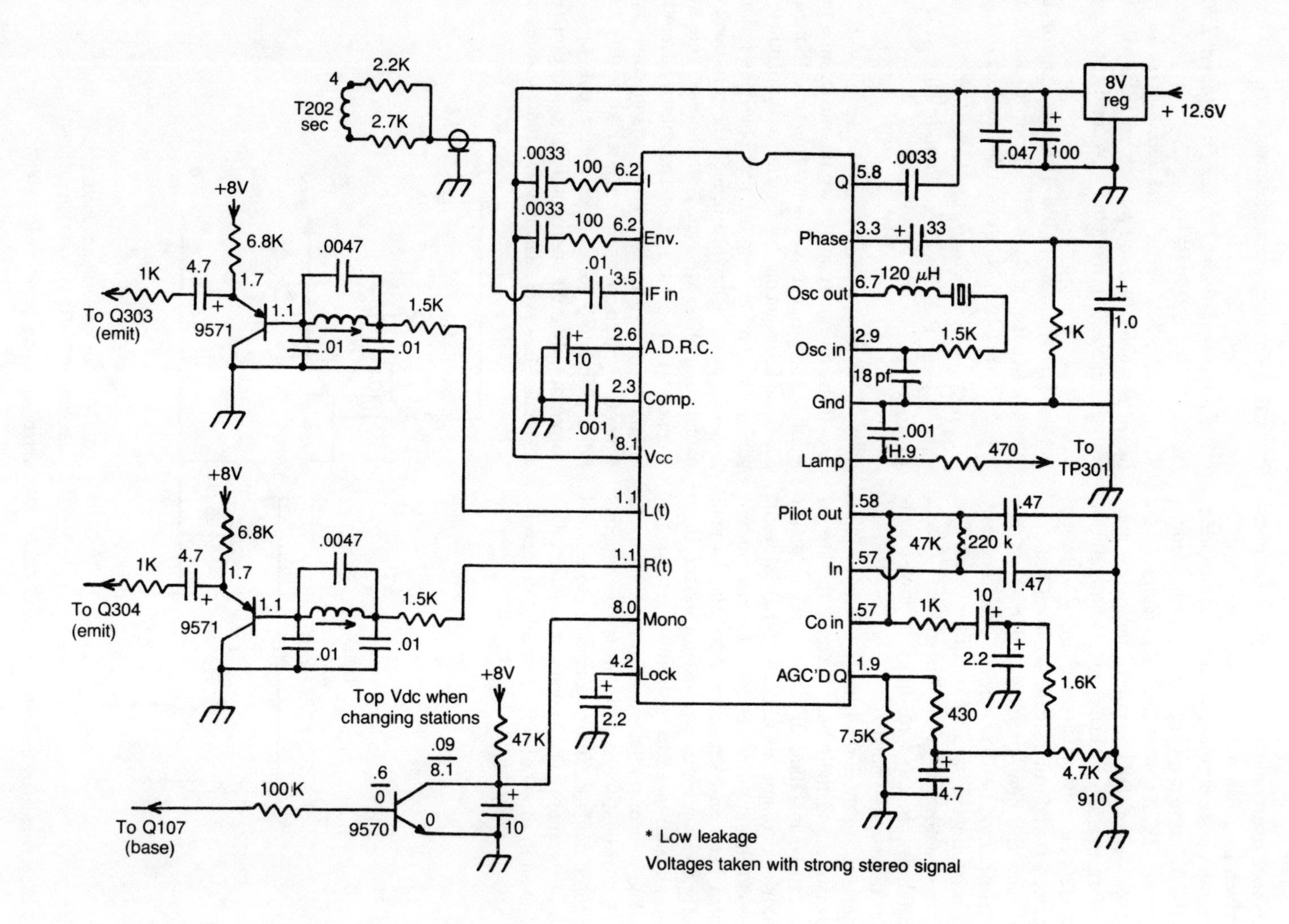

Fig. 5-18. The new AM stereo decoder is installed (courtesy Motorola).

of its original broadcast station-monitor design and was modified by Motorola to include a dual bandwidth IF panel and high performance decoder illustrated in Fig. 5-19. The complete conversion is shown in Fig. 5-20. Using the MC13020, typical distortion expected is 1 percent maximum with a minimum separation of 28 dB.

In the dual bandwidth modification, transformer T1 couples and loads the wideband filter (less capacitor C2). The filter element is identified as a 6-pole Murata SFH 450D, with a minimum 6 dB bandwidth of ±10 kHz. R4 and C3 optimize the noise characteristic filter response. Diodes D1 through D3 bias gate sources of the two dual gate MOSFETs at 21V, which are used as wide and narrowband switches.

Transformer T2 is the leading and coupling transformer for the Toko 450 DFMS narrowband filter, having a 6 dB bandwidth of ±4.5 kHz. R16 and C6 optimize filter "noise" response. The overall circuit is designed to work in series deriving benefit from both filters at narrow bandwidth.

The input to this filter network reaches the receiver through the emitter of Q102, and its output returns to IC Q101 through pin 9. Decoder card input also devices from IC Q101, and the card which decodes C-QUAM as well as the pilot tone detector contains the following:

- ☐ An IF amplifier to adjust input levels to 200 mV rms (pot R2).
- ☐ A pilot tone center frequency adjustment (pot R34).
- ☐ A pilot tone IM notch filter due to Q5 and associated circuits.
- ☐ A L and R channel separation adjustment (R45, R53).
- ☐ An adjustable 10 kHz notch whistle filter for sideband (R62, R71).
- ☐ L1 and L2 for narrow bandwidth filtering and 10 kHz adjustable notch.
- ☐ Output buffering to drive most audio preamplifier inputs.

The sideband lowpass filter in IC1 is added to smooth out the transition between an IF flat-nosed response and a rapid attenuation of the skirt responses to avoid any audio ringing and sound balance improvement. Potentiometer R7 adjusts the dc value of the envelope detector to that of the I detector, and R8 adjusts the Q detector for phase offset, but is normally not necessary.

The MC13020, naturally is the initial left and right output intelligence supplier and routes signals through the various IC1 amplifiers to the MC14529 buffer, which then does its data selection upon signal receipt and/or switch execution. This is a 12 Vdc operated IC, the voltage swings can be fairly large and a substantial signal may be expected at the inputs of IC4, the final channel amplifiers.

The MC13020P we've already described at some length, so the multitude of surrounding passive biasing, and series filter components should come as no surprise. Nor should the usual audio capacitative coupling have any significance since it's common practice and an easier design method at these low audio frequencies. Both IC1 and IC2 are 14-pin package amplifiers with multiple sections and in various configurations. There are a few potentiometer adjustments and lots of resistive biasing, indicating restricted current flow among MC13020 feeds, but somewhat more push with the IC1 group supplying MC14529B.

That should pretty well sum up Motorola's forthright contribution to this chapter's material. Apparently they're making a major effort to outdistance competition as rapidly as possible, and nothing promotes success like hard work.

A Harris Stereo Decoder

Once again David Hershberger of Harris' Broadcast Division devotes his prime talents and engineering proficiency to aiding many who would convert their AM monophonic radios to AM stereo, particularly for the Harris linear system. But be advised that not all AM radios will convert—especially the inexpensive ones—and only those with low spurious local oscillator FM or PM and low incidental phase modulation on strong signals are viable candidates. Usually, also, electronically tuned radios have few or no microphonics and are better suited to conversion than mechanically tuned sets.

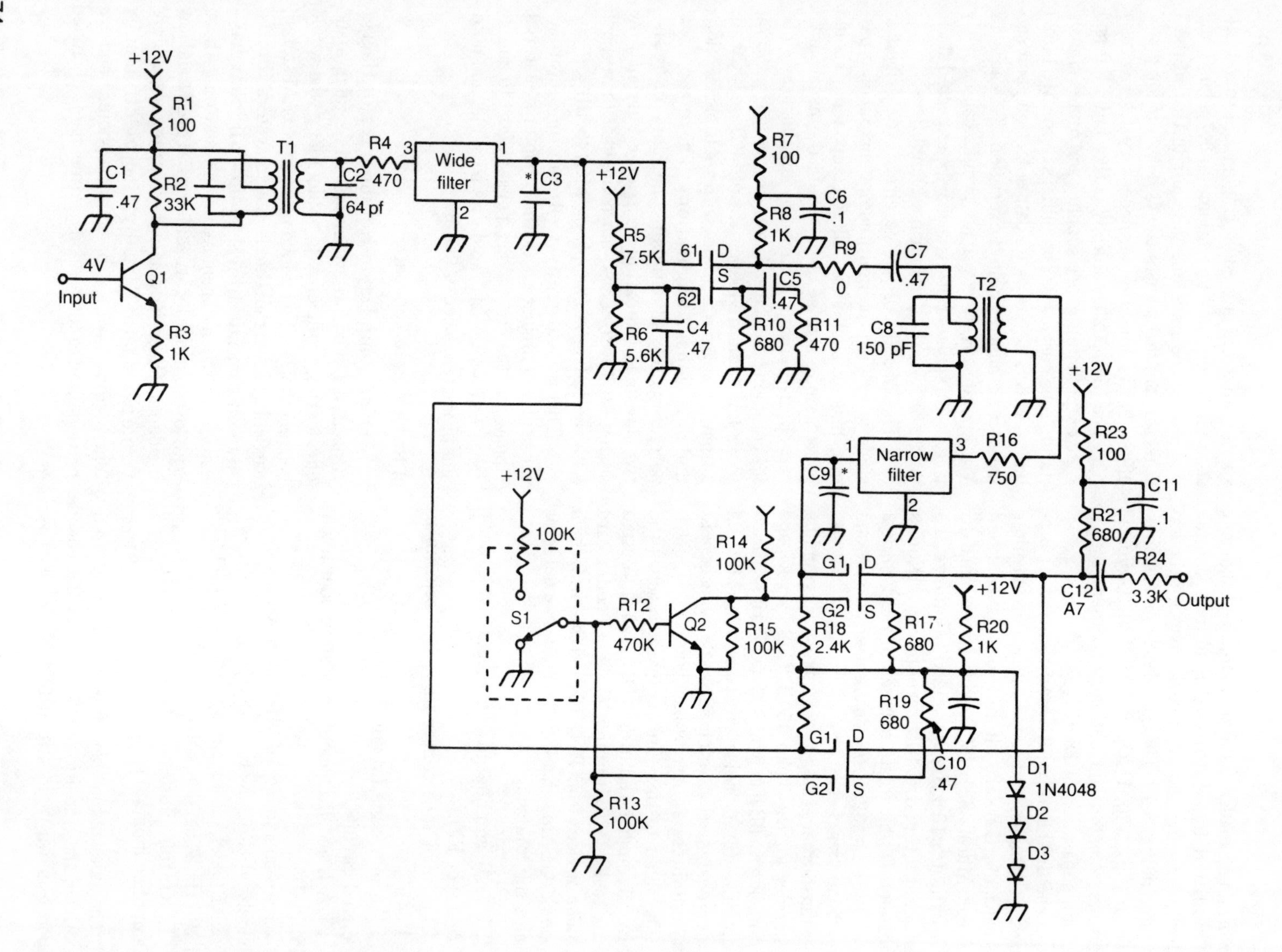

Fig. 5-19. Adding the wideband/narrowband switch function for SX-6 (courtesy Motorola).

So before you begin, better set up a few tests to determine if your particular radio passes or fails. Hershberger suggests you go about it this way: Find a beat frequency oscillator (bfo), tune the radio somewhere in a quiet portion of the AM band, generate a CW signal, then inject bfo into the IFs or other convenient point. Resulting audio should be clean with no warbling or pitch changes that would indicate spurious FM. Low incidental phase modulation is next, and this is particularly significant for electronically tuned radios receiving strong signals. Varactor capacitance chanes with varying rf signals and induces incidental nonlinear phase modulation which, according to Mr. Hershberger, should be less than 10 degrees for satisfactory stereo reproduction.

After you've patched the appropriate system together, you can measure incidental quadrature modulation by tuning in a monophonic station, displaying I (in-phase) and Q (quadrature) outputs of the decoder on a fairly well phase-matched X-Y oscilloscope that uses equivalent Ch. 1 and Ch. 2 inputs for the CRT display. The display should appear as a straight line with amplitude representing modulation.

Since the Harris AM stereo design requires no high order sidebands, IF filtering is no different than for good mono. You will also find that synchronous detection does not require precise station tuning since, unlike a diode envelope detector, it is linear and regenerates its own carrier signal as a switching reference to discriminate against noise and distortion.

Receiver Modifications

If these tests don't disqualify your radio, you'll need to bring out an IF signal of 200 mV rms (unmodulated carrier), preferably from a source such as an emitter follower. You'll also need a reasonably stiff AGC whose time constant is slow enough to not distort 50 Hz envelope modulation. You're in trouble if you can't reach the IF AGC. But if AGC voltage is simply low, then use an amplifier for better gain-response. Extract a Vcc of from +12 to 20 operating volts to be made available from a regulated power supply.

The stereo decoder will use a Motorola MC13020P integrated circuit (already discussed separately and in detail) for the synchronous detector in both sum and difference channels. But in order to use it, the pilot frequency must be moved to 55 Hz (now 25 Hz), making threshold equivalent to 7.5 percent pilot injection. Co-channel lockout sensitivity may be decreased, the distortion correction function (error amplifier) is then disabled by a large capacitor so L and R outputs develop envelope plus quadrature and envelope minus quadrature at pins 7 and 8 (Fig. 5-21), and the difference is the quadrature or L−R channel. Simple highpass filters remove the pilot.

The L+R signal is next. In-phase and envelope information appear at pins 1 and 2. A FET switch conducts the I channel (L+R) when the phase-lock loop stabilizes. When not locked, the mono sum signal develops from the envelope detector at pin 2 and is routed to phase shifter Q5, matching the phase shift in L−R. Q1-Q2 consist of a JFET analog switch controlled by PLL lock from pin 10. Here Hershberger warns that these FETs have a low 3-volt pinchoff voltage and any substitutes should have a pinchoff of at least 5V or less. The Q3-Q4 bipolars below, of course, form differential triggers for the FETs.

The sum signals now pass to dual "dematrixing" operational amplifiers IC2 which combine these with L−R coming out o of IC1 and the phase shifting networks into the opamps. Note there are two oscillator circuits available for use with electronic or mechanical tuning. One is an L/C type (drawn separately) and the other a crystal-controlled type preferred with electronic timers since it "provides better immunity from mechanical shock."

For those using the L/C oscillator, tune in a mono or stereo station and carefully adjust for adequate center tuning and positive lock in. This occurs at about 4 volts measured at pin 19. Mr. Hershberger warns that some radios, due to PM noise or incidental phase modulation in the local oscillator, will trip stereo lockout so you may have to flick the auto/mono stereo switch to mono and then back again for the stereo light to appear. If so,

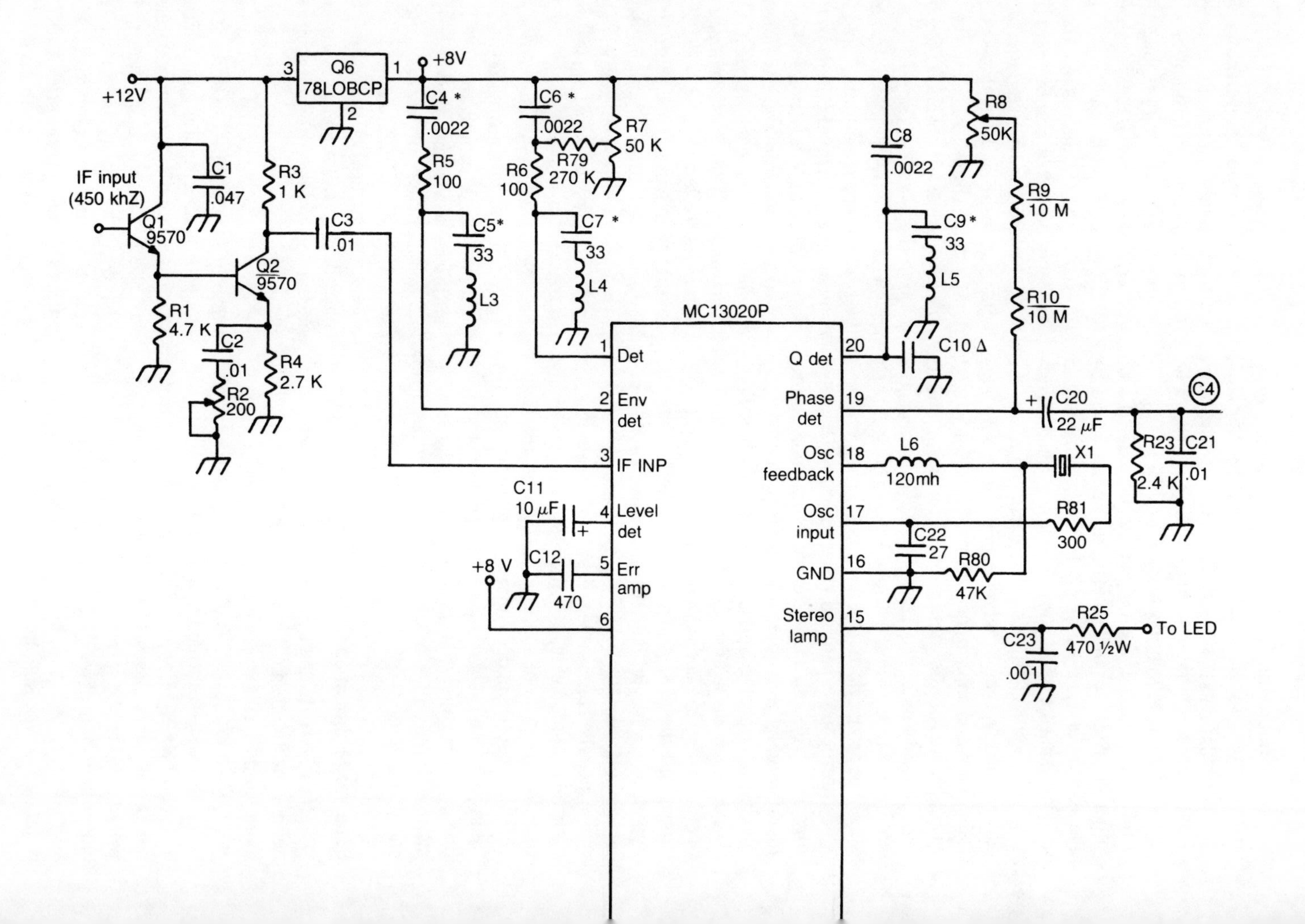
+12V
IF input
(450 khZ)
Q1
9570
C1
.047
R1
4.7 K
Q6
78LOBCP
3
1
2
+8V
R3
1 K
C3
.01
Q2
9570
C2
.01
R2
200
R4
2.7 K
C4 *
.0022
R5
100
C5*
33
L3
C6 *
.0022
R7
50 K
R79
270 K
R6
100
C7 *
33
L4
C8
.0022
R8
50K
R9
10 M
C9 *
33
L5
R10
10 M
MC13020P
1 Det
2 Env det
3 IF INP
4 Level det
5 Err amp
6
20 Q det
19 Phase det
18 Osc feedback
17 Osc input
16 GND
15 Stereo lamp
C10 Δ
C20
22 μF
C4
R23
2.4 K
C21
.01
L6
120mh
X1
R81
300
C22
27
R80
47K
C11
10 μF
+8 V
C12
470
C23
.001
R25
470 ½W
To LED

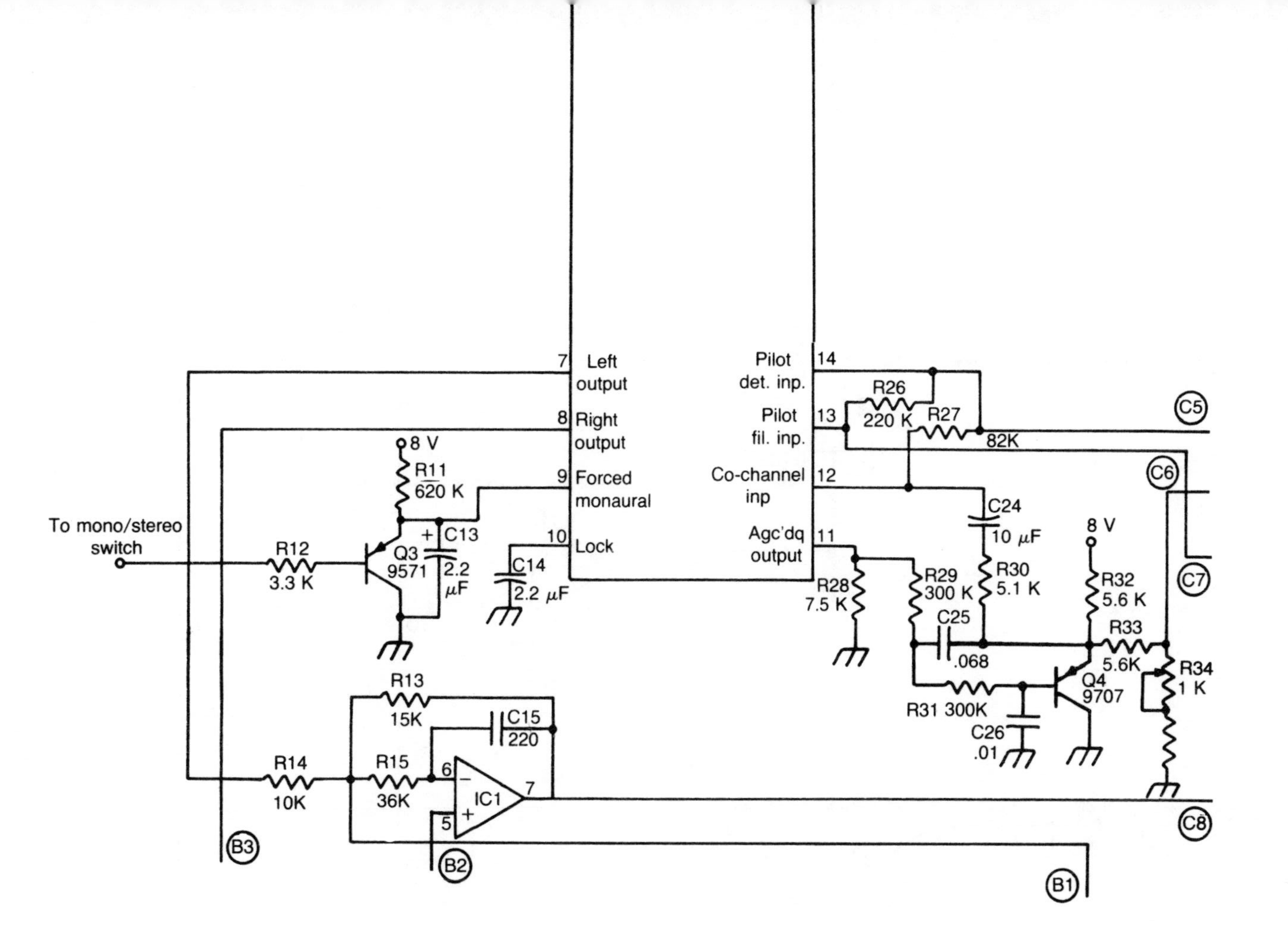

Fig. 5-20. The actual SX-6 AM stereo conversion (courtesy Motorola). (Continues through page 80.)

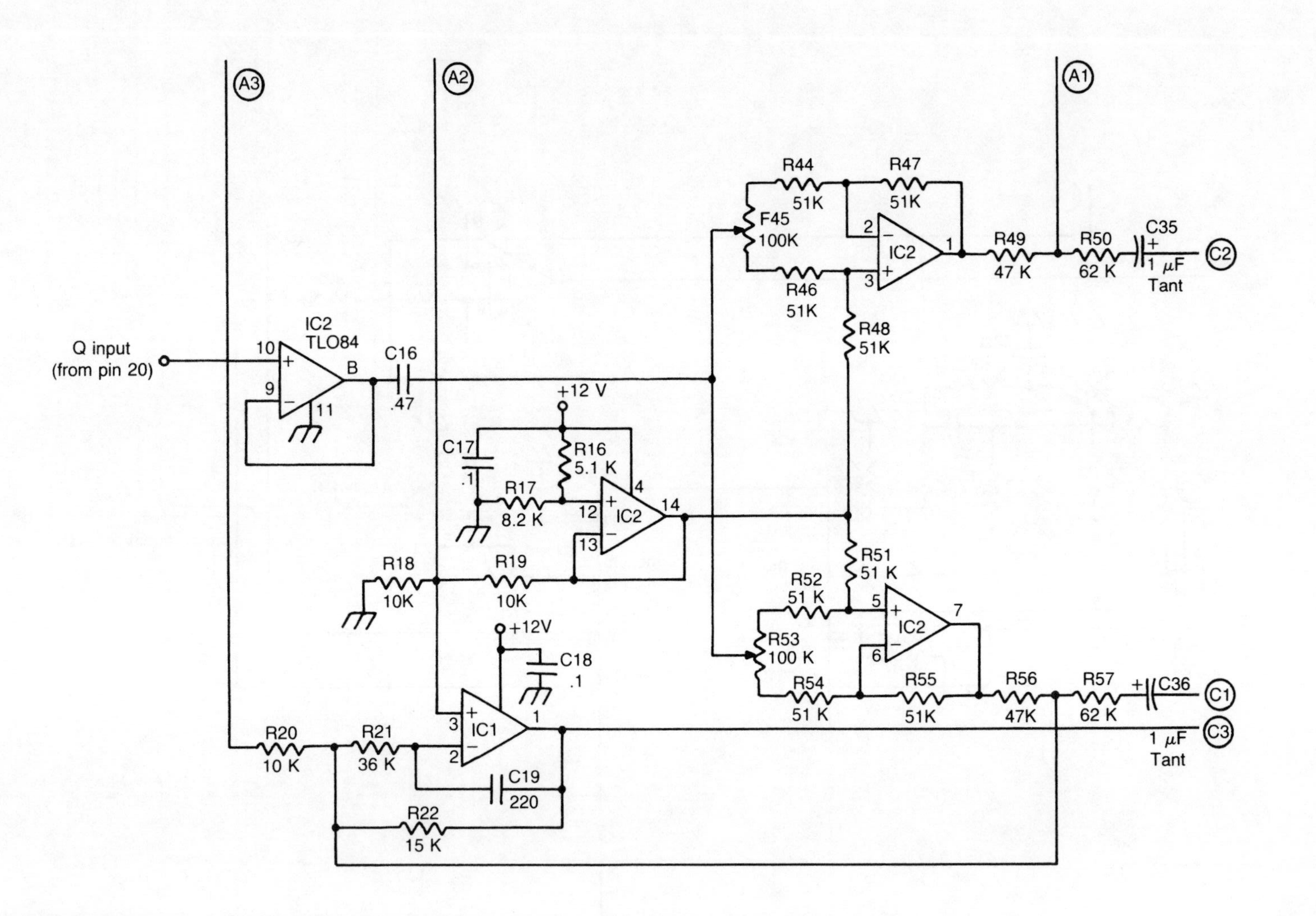
A3
A2
A1
R44
51K
R47
51K
F45
100K
IC2
R49
47 K
R50
62 K
C35
1 µF
Tant
C2
R46
51K
R48
51K
Q input
(from pin 20)
IC2
TLO84
B
C16
.47
+12 V
C17
.1
R16
5.1 K
R17
8.2 K
IC2
R18
10K
R19
10K
R51
51 K
R52
51 K
IC2
R53
100 K
+12V
C18
.1
R54
51 K
R55
51K
R56
47K
R57
62 K
C36
C1
C3
1 µF
Tant
IC1
R20
10 K
R21
36 K
C19
220
R22
15 K

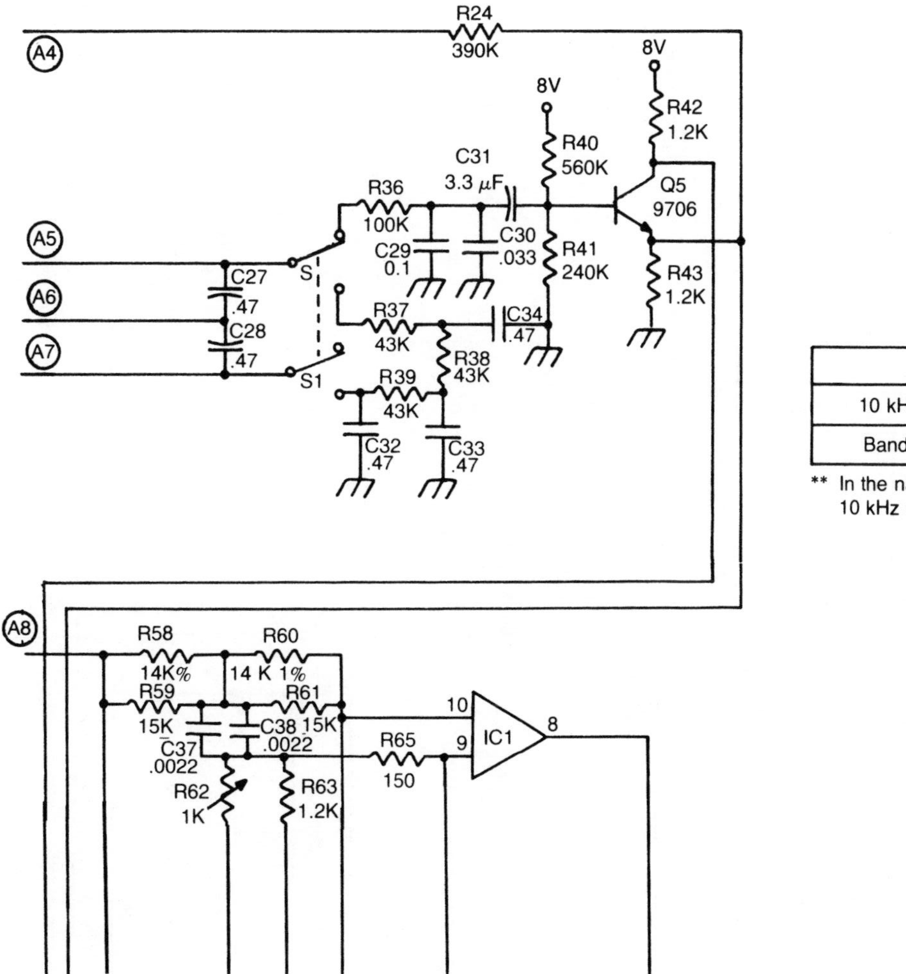

	"0"	"1"
10 kHz notch filter	No notch	Notch
Bandwidth switch	Wide	Narrow**

** In the narrow bandwidth position, 10 kHz notch filter is always *on*.

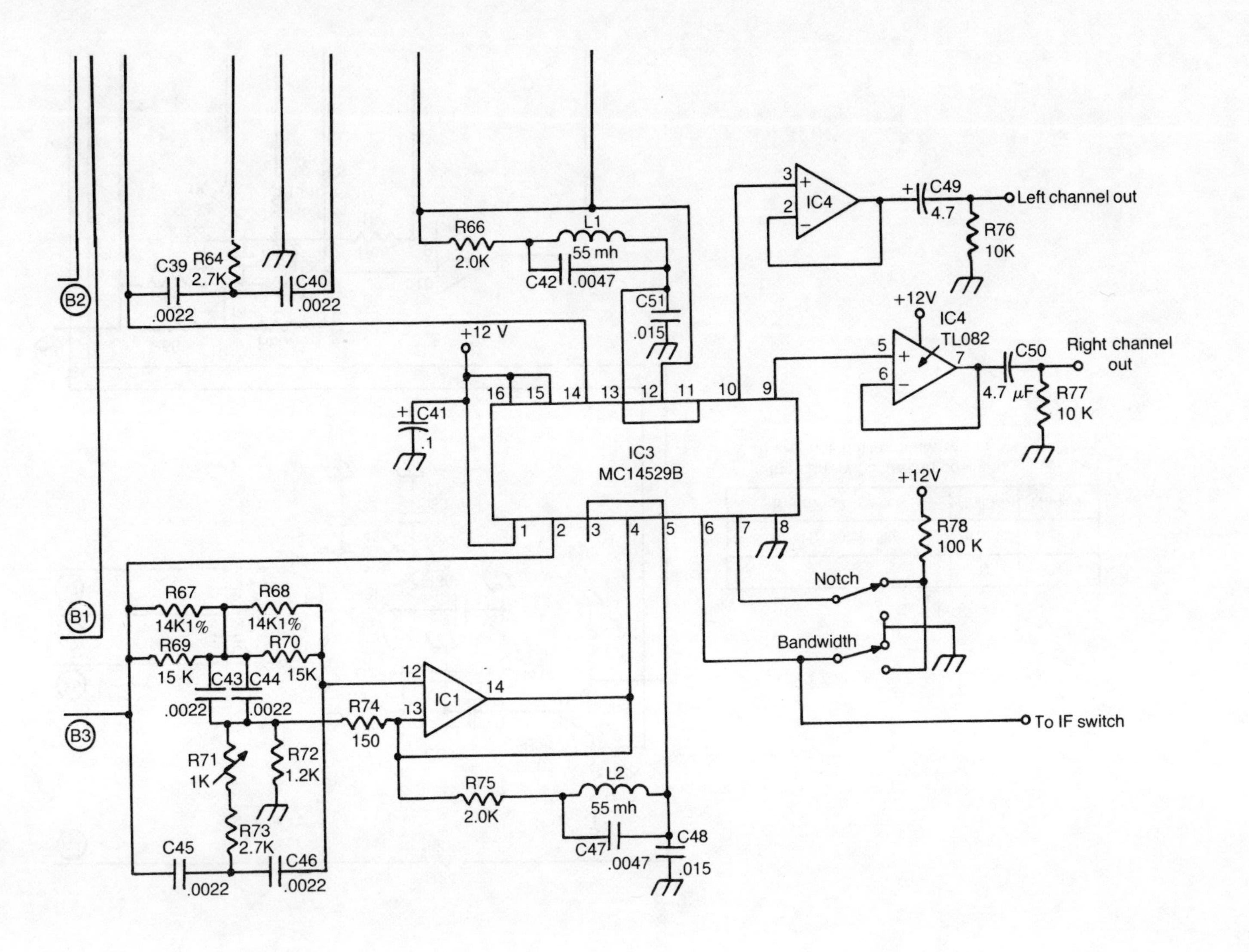
IC4
Left channel out
C49
4.7
R76
10K
+12V
IC4
TL082
C50
Right channel out
4.7 μF
R77
10 K
R66
2.0K
L1
55 mh
C42
.0047
C51
.015
C39
.0022
R64
2.7K
C40
.0022
+12 V
C41
.1
IC3
MC14529B
+12V
R78
100 K
Notch
Bandwidth
To IF switch
R67
14K1%
R68
14K1%
R69
15 K
R70
15K
C43
.0022
C44
.0022
IC1
R74
150
R71
1K
R72
1.2K
R73
2.7K
C45
.0022
C46
.0022
R75
2.0K
L2
55 mh
C47
.0047
C48
.015
B1
B2
B3

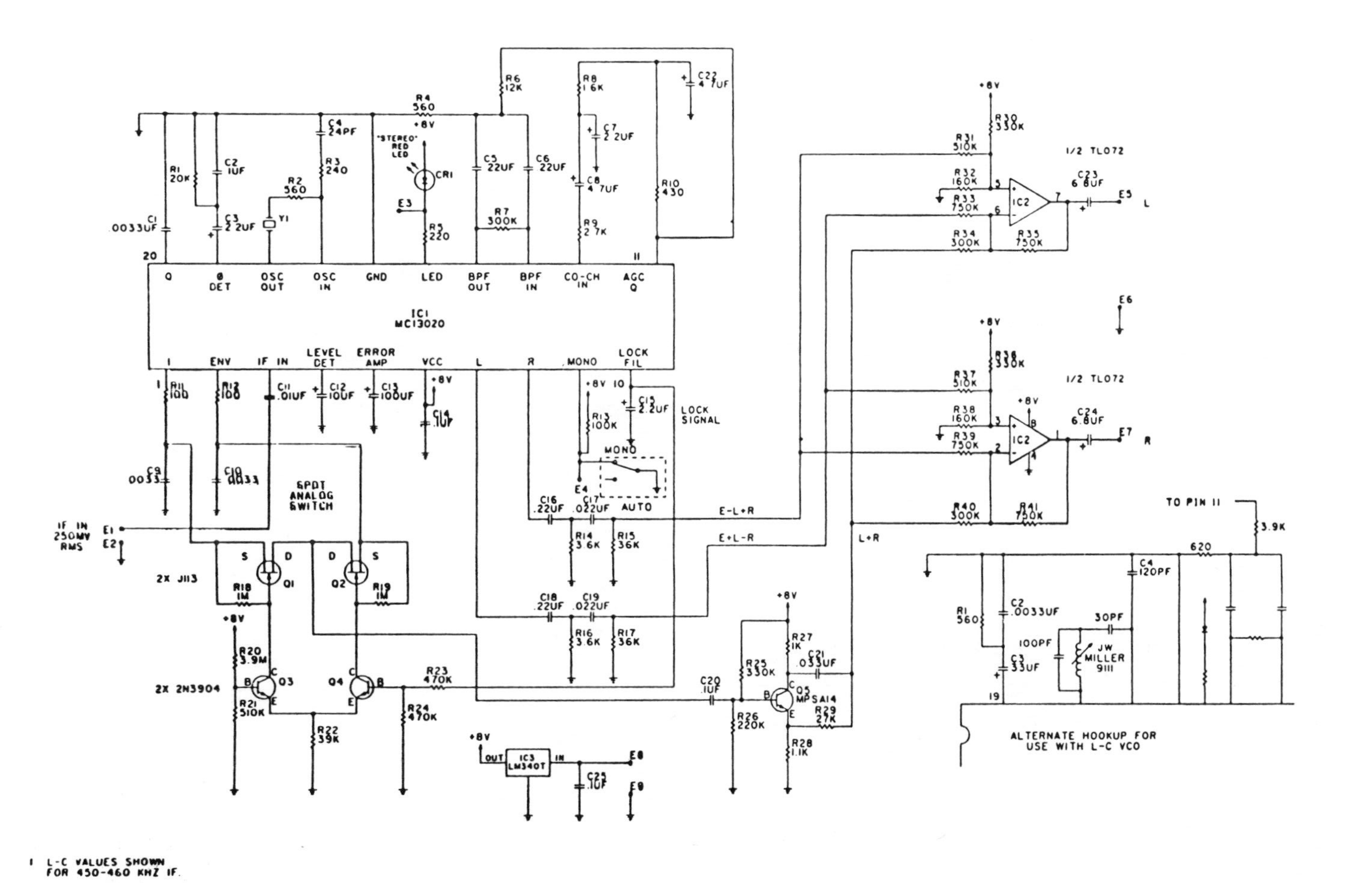

Fig. 5-21. Harris' AM stereo decoder (courtesy Harris Corp.).

you'll have to increase the 2.2 μF capacitor to 47 μF and/or remove the 2.7K resistor at pin 12, leaving an open circuit.

Following Mr. Hershberger's instructions, that's about the extent of this modification which probably works satisfactorily with many of the better AM radios. His recommended parts list is Table 5-1.

DELCO RADIO

General Motors radio division, Delco Electronics, has now "adopted" the Motorola C-QUAM system and is going ahead full bore with AM/FM stereo radios for Buicks (initially) in the 1985 model year. It may also be installed in certain 1984 models of Chevrolet, Pontiac, Oldsmobile, Buick, Cadillac, and even some GMC/Chevy trucks. Factory installation UP2 applies to Buick Century only for now.

with a frequency synthesizer clocked at 50 Hz serving both AM and FM tuning as well as the microprocessor, which is the real controller of the entire system. For, in addition to time set, user, and AM/FM control functions, this microprocessor also manages tape/radio mute, headlamp dimmer, and external air controls.

Observe that the FM tuner has keyed AGC, and the standard IF AGC, while only the second AM rf has AGC emanating from the threshold adjust and bandwidth switch. Note also that DM-235 has a 3.601 MHz crystal-controlled oscillator, suggesting tightly controlled AM stereo detection which is subsequently 10 kHz whistle-filtered before reaching the mute control and stereo separation circuits. After the mute (cutoff) IC, outputs are further amplified and routed to the separate speakers for audible reception.

The FM section, according to the diagram,

Table 5-1. Parts List.

Part	Description
C1	.0033 mF ceramic disc or mylar
C2	.1 mF Mylar*
	.0033 mF Mylar**
C3	2.2 mF tantalum or electrolytic*
	33 mF tantalum or electrolytic**
C4	24 pF mica*
	120 pF mica**
C5	.22 mF Mylar
C6	.22 mF Mylar
C7	2.2 mF tantalum or electrolytic
	(use 47 mF to decrease stereo lockout)
C8	4.7 mF tantalum or electrolytic
C9	.0033 mF ceramic disc or Mylar
C10	.0033 mF ceramic disc or Mylar
C11	.01 mF ceramic disc
C12	10 mF tantalum or electrolytic
C13	100 mF tantalum or electrolytic
C14	30 pF**
	deleted*
C15	2.2 mF tantalum or electrolytic
C16	.22 mF Mylar
C17	.022 mF Mylar
C18	.22 mF Mylar
C19	.022 mF Mylar
C20	.1 mF Mylar
C21	.033 mF Mylar
C22	4.7 mF tantalum or electrolytic
C23	6.8 mF tantalum or electrolytic
C24	6.8 mF tantalum or electrolytic
C25	.1 mF ceramic or Mylar
C26	100 pF**
	deleted*
CR1	Red LED (Stereo indicator)
IC1	MC13020P (Motorola)
IC2	TL072 or 4558 dual opamp
IC3	LM340T-8 or 7808 +8 volt regulator
L1	Adjustable Coil**
	J.W. Miller #9111
	12.5-29.45 uH
	Nominal value 15.5 uH
	deleted*
Q1,Q2	J113 N-channel switching FET (3 volt pinchoff)
Q3,Q4	2N3904 general purpose NPN
Q5	MPSA14 Darlington NPN
R1	20 k ohm*
	560 ohm**
R2	560 ohm*
	deleted**
R3	240 ohm*
	deleted**
R4	560 ohm**
	620 ohm**
R5	220 ohm
R6	12 k ohm*
	3.9 k ohm**
R7	300 k ohm
R8	1.6 k ohm
R9	2.7 k ohm
R10	430 ohm
R13	100 k ohm
R14	3.6 k ohm
R15	36 k ohm
R16	3.6 k ohm
R17	36 k ohm
R18	1 Megohm
R19	1 Megohm
R20	3.9 Megohm
R21	510 k ohm
R22	39 k ohm
R23	470 k ohm
R24	470 k ohm
R25	330 k ohm
R26	220 k ohm
R27	1 k ohm
R28	1.1 k ohm
R29	27 k ohm
R30	330 k ohm
R31	510 k ohm
R32	160 k ohm
R33	750 k ohm
R34	300 k ohm
R35	750 k ohm
R36	330 k ohm
R37	510 k ohm
R38	160 k ohm
R39	750 k ohm
R40	300 k ohm
R41	750 k ohm
Y1	Quartz crystal*
	8 times IF frequency
	parallel resonant, 20 pF load capacitance
	(example: 450 KHz IF frequency requires 3.6 MHz crystal)
	deleted**

* denotes parts for ETR crystal controlled version
** denotes parts for L-C oscillator version

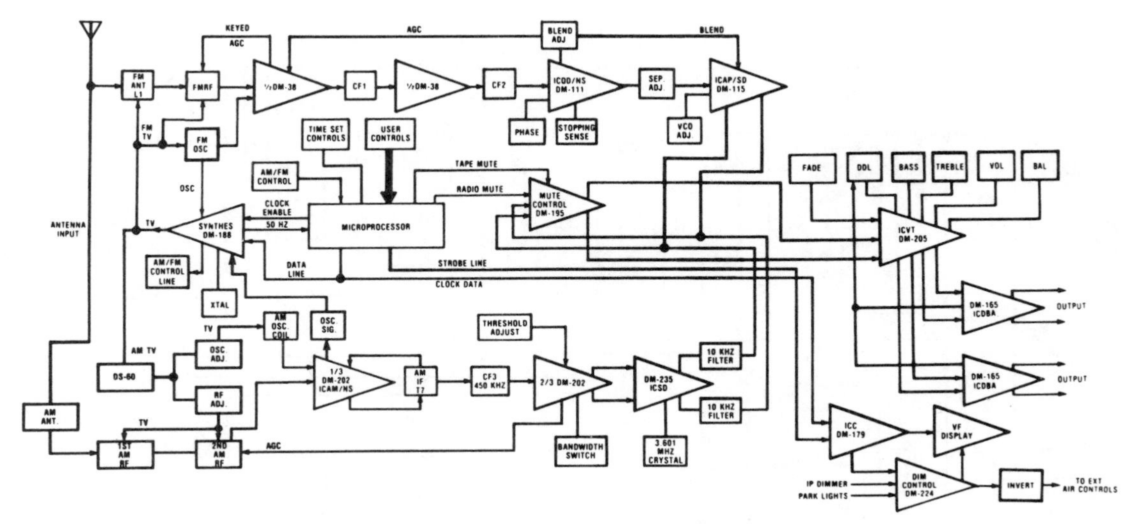

Fig. 5-22. Block diagram of Delco's AM/FM full stereo auto radio (courtesy Delco Electronics).

Features are: tuner station scan and seek (electronically), preset four AM and four FM stations, digital clock and frequency display, switchable AM bandwidth for day/night reception, balance, tone, treble, bass and monophonic controls.

Delco, in its 27D-1984-2 M service bulletin, prints both a block diagram and schematics of its new radios, identifying them as the 2000 series. The block diagram (Fig. 5-22) is fairly compact and will be displayed in full. Schematics of receiver, audio control, logic and display boards are quite large, so only the receiver board layout will be printed. Since there is *no* overall engineering writeup, we'll wade through as best we can, beginning with the block diagram. Parts lists consist of the usual discrete transistors, diodes, varactors, capacitors, filters, controls, transformers, and switches, but it also includes 18 interesting modules and three RC networks. These modules contain such electronics as comparators, synthesizers, quadrature detector, AM stereo detector, AM IC noise suppressor and, of course, all the various integrated circuits. Letters DM identify the group modules.

In the block diagram, antenna inputs go to both the FM (top) and AM (bottom) receiver electronics. As usual, there are both AM and FM oscillators, but appears conventional except for a circuit called a "blend adjust" which we may be able to understand after persuing the schematic. Controls consist of Fade, Bass, Treble, Volume, and Balance, along with a circuit called Dynamic Distortion Limiting (DDL) that could be some type of automatic compression or outright clipping when levels exceed normal design limits. As you might expect, DM-205 and DM-165 are located on the audio control board, which we don't show schematically since these are simply outputs that are completely dependent on signal processing by the remainder of the system.

The receiver board schematic (Fig. 5-23) contains six ICs, 13 transistors, a few diodes, some coils, and lots of capacitors and resistors. FET and dual channel MOSFET transistors receive AM and FM signals respectively through tuned circuits and process them through the usual series of amplifiers and transformers to FM and AM mixers which reduce these signals to 10.7 MHz and 450 kHz IFs, respectively. FM then continues through IC amplifiers and the various filters, an LC choke circuit and RC coupling, where it is detected and further amplified for right and left audio outputs.

Both AM and FM oscillators are inductor core-adjustable, and there are potentiometer rf and oscillator adjusts in AM, as well as a second rf

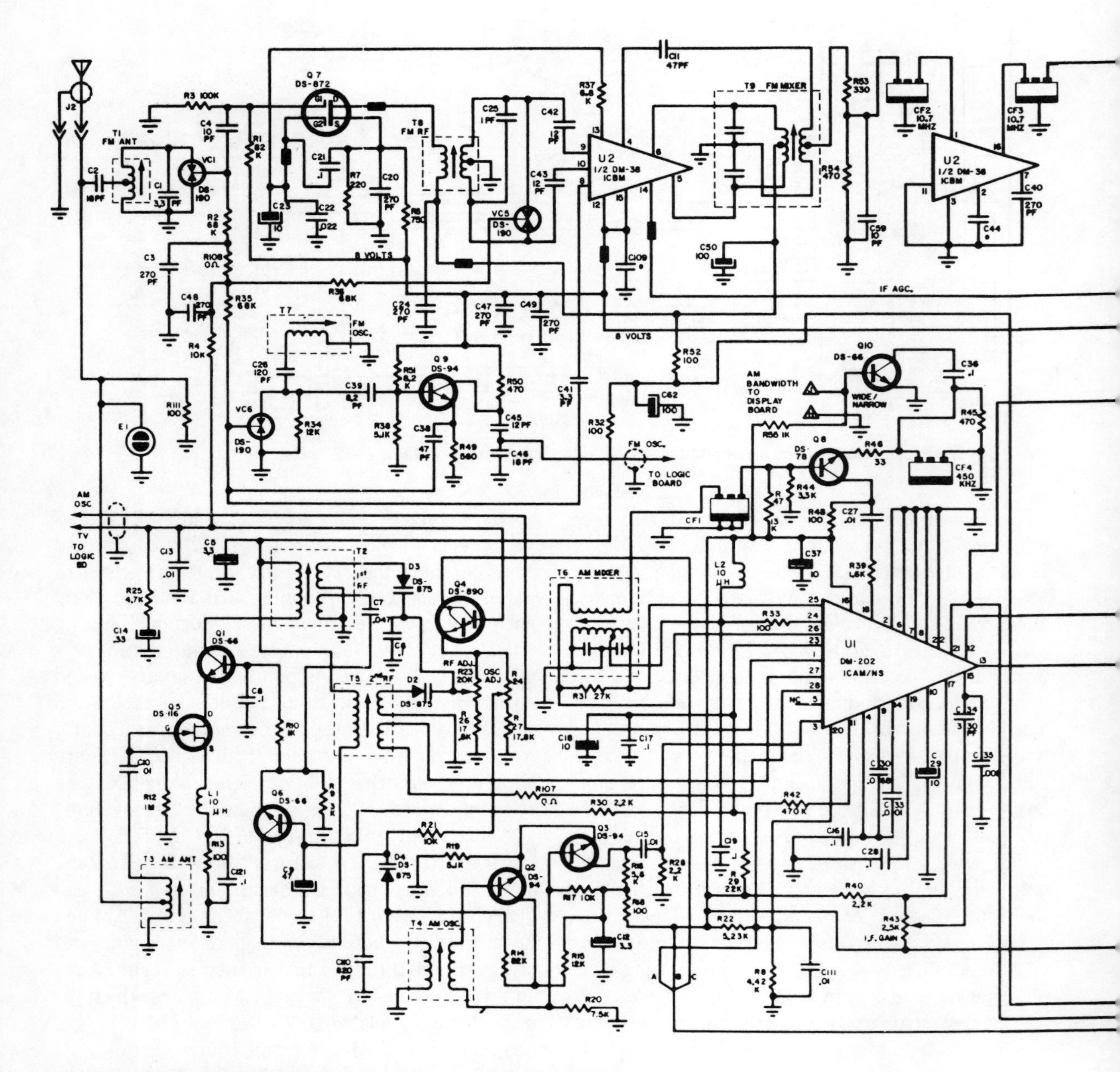

Fig. 5-23. Schematic of Delco's receiver board (courtesy Delco Electronics).

transformer. DM-202 then combines these signals for the 450 kHz IF whose gain is also potentiometer adjustable. AM detection occurs in DM-235 and right and left channels are 10 kHz whistle-filtered for the 9-pin connector to the logic board and thence to the audio control board with its bass, fade, treble adjustments and final power amplifiers. The curious "blend adjust" amounts to a potentiometer output-

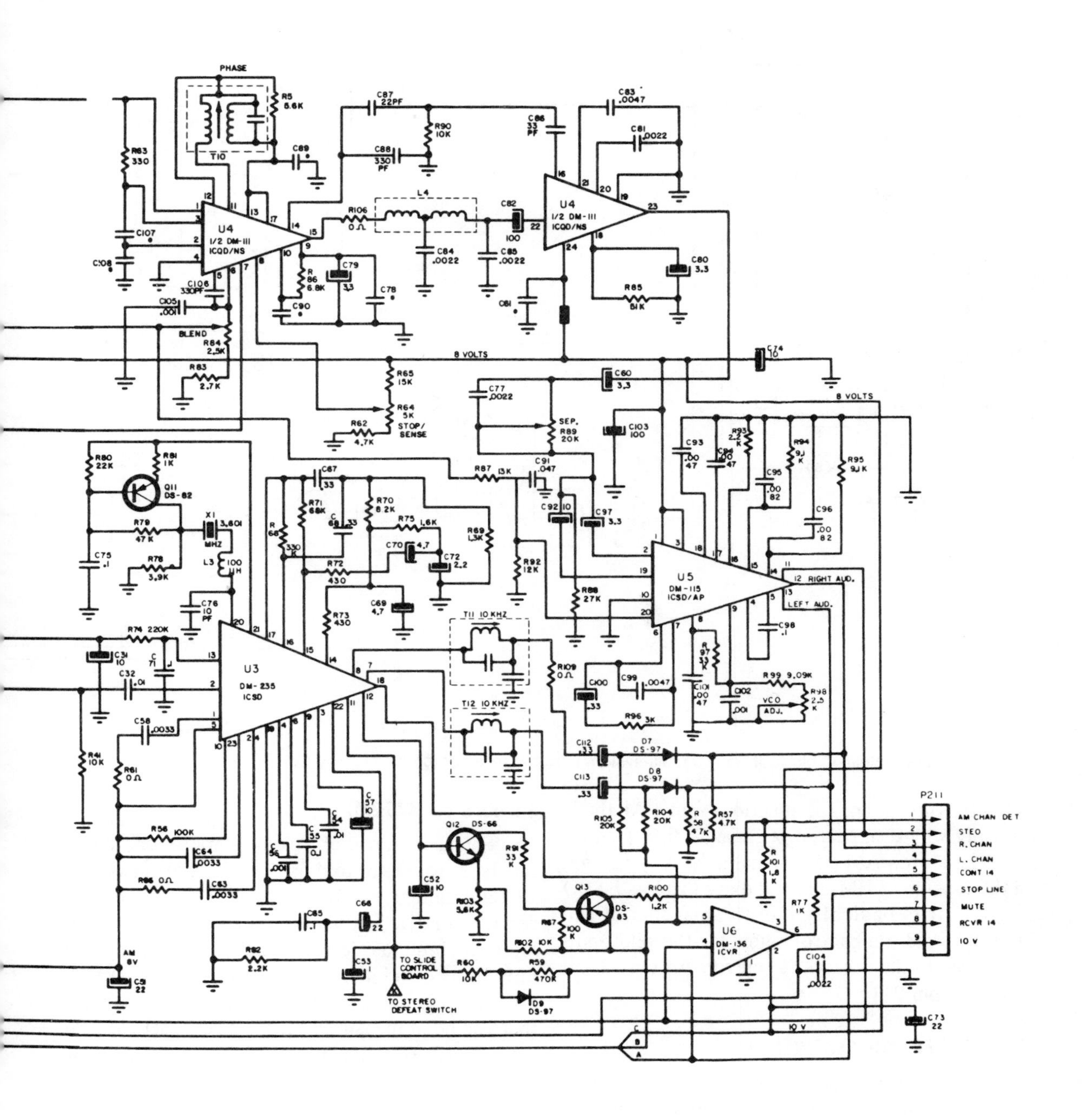

ting from DM-111 as part of the IF AGC line and continuing on to right and left output module DM-115. Here it applies a well-filtered dc voltage that must rise and fall with signal strength probably maintaining an equitable L/R "blend" for best stereo reception—at least that's what the circuit appears to do from its electrical connections since the arm of any potentiometer is its normal output.

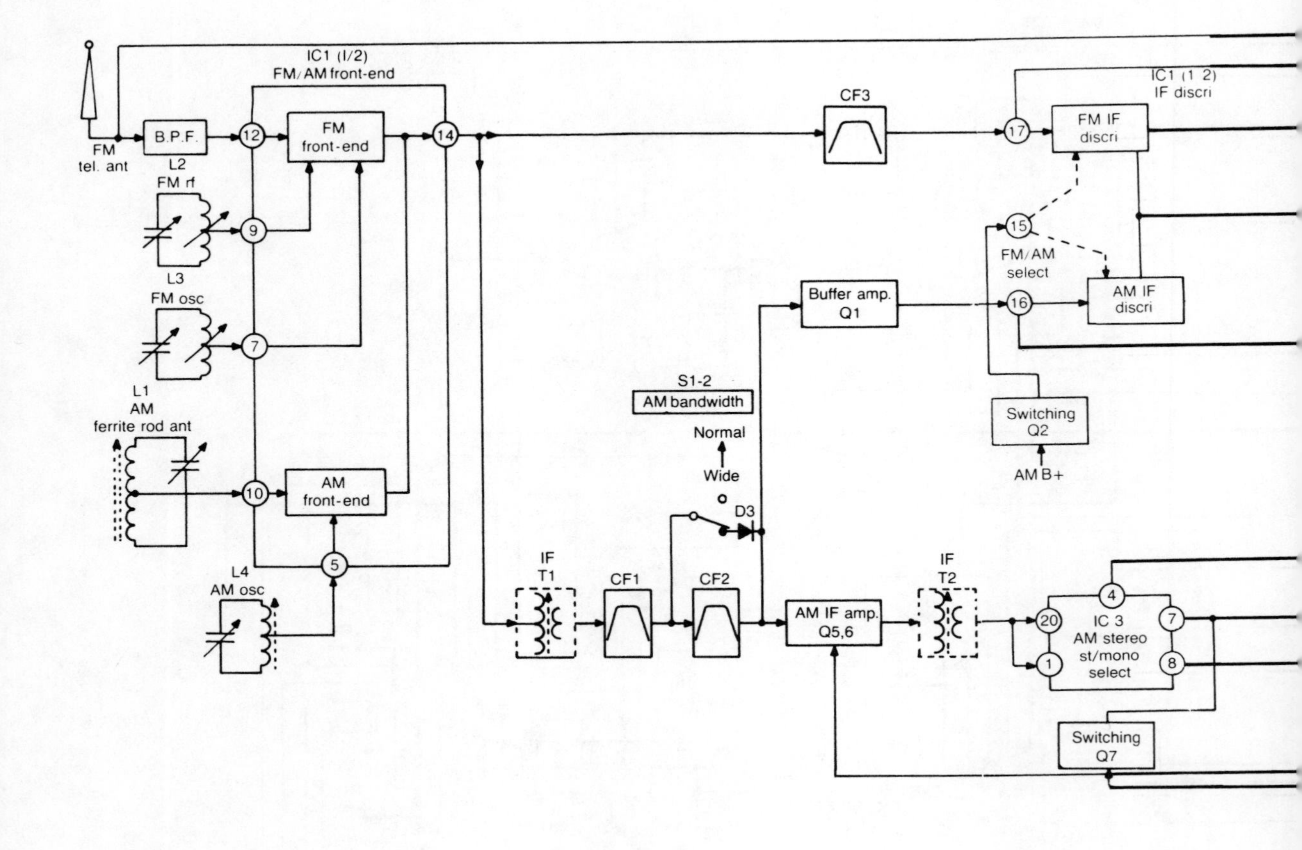

Fig. 5-24. Highly simplified block of Sony's multi-system portable receiver (courtesy Sony Corp.)

Actually, this is an FM function—not AM that's probably why it appeared peculiar at first glance.

The stereo AM pilot signal is heavily filtered on the topside of DM-235 and exits P211 as STEO at pin 2. It is basically identical to that of the MC13020 and the control signal for the pilot indicator goes to the receiver's controls via pin 18 and STEO on pin 211.

MULTI-RECEPTION RECEIVERS

While there's incomplete information on this subject, a good deal of R & D development is concentrated on 4-system decoder ICs, two of which are noted and explained with the information available.

Sansui's TU-S772MX

Sansui Electric of Tokyo, markets one home receiver that we're told has 11 ICs, 22 transistors, 12 potentiometers, 108 resistors, 45 capacitors, and 13 overall adjustments. Offering a phase-locked loop synchronous detector that is said to reduce interference and distortion, the new set reputedly provides a frequency range of from 50 to 15,000 Hz (but has been measured at less than that) and will receive any of the four AM stereo systems now existing. In later versions due for 1985 production, a number of these transistors are scheduled to be supplanted by one or more integrated circuits, with probably substantial improvements over previous discrete versions. Competitors complain the early

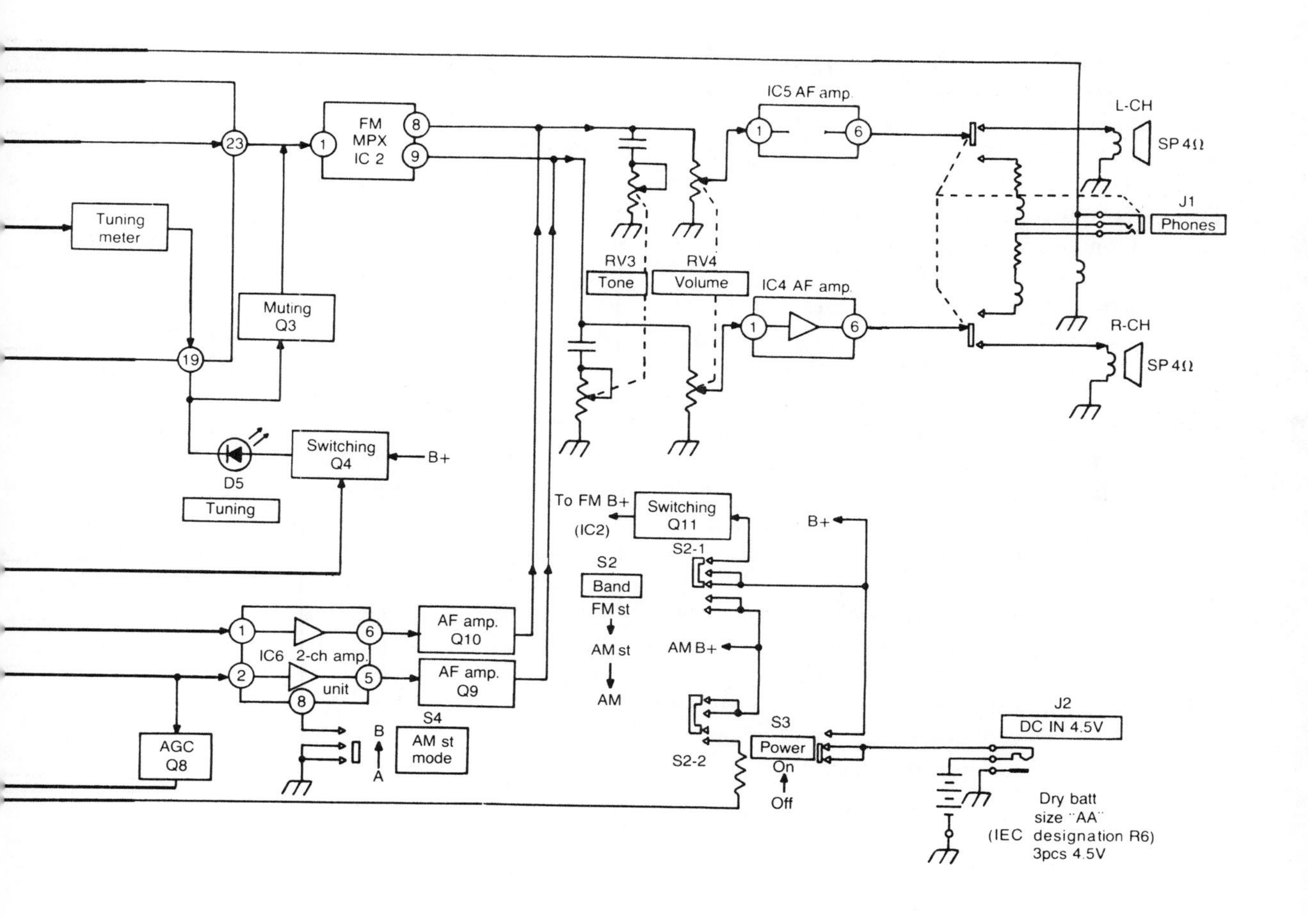

models were subject to "falsing" (incorrect mono/ stereo switching), were not good weak-signal performers, and occasionally took 20-30 seconds to recognize a stereo signal. Unfortunately, there's no further technical information on the TU-S77AMX available at this time—and probably not until early 1985. We're told that basically the set is a Harris decoder with a 90° phase shift for Kahn.

Sony's Portables

Sony Consumer Products already has a pair of portables available (including Walkman), with one year parts and labor guarantee. Later, parent Sony Corp. of Japan expects to produce (or make available) automobile radios, followed by standard AM/FM home stereo models. The SRF-A100 version comes fitted with shoulder strap, AM bandwidth selector (wide or narrow), phone jack, all-type reception, and operates on 3 AA batteries, or 120 Vac with ac adaptor. Weight (with batteries) is only 1½ lbs.

As usual, the Japanese are coy with any technical information, but those familiar with the two models of receivers say the 4-way decoders (with flip switch for Kahn) offer very reasonable sound and respectable signal sensitivity. A pair of 3-inch dynamic speakers provide stereo sound, or headphones may be used for privacy.

A block diagram of one unit (Fig. 5-24) gives an idea of what the general system is all about. If the count is accurate, there are 6 integrated circuits, the usual coils, and a dozen or more discrete transistors Block waveforms below CR1-3 probably designate passband filters, with bandwidth switch S1-2 either shunting CF2 for narrow band operation, or permitting it to rein in the circuit for wideband rsponsm gnote that the AM and IF discriminators are all contained in IC1, as are the FM/AM front ends. IC3 and asociated transistors furnish. AM mono switching, and there's a tuning meter with LED indicator for best station selection. Left and right AM outputs join FM multiplex IC2 for final amplification by ICs 4 and 5 for either phone jacks or 4-ohm twin speakers. You'll also find the usual bandswitch for AM/FM stereo or monophonic, plus volume, tone controls, and the usual station tuning. Antennas are a whip for FM and a ferrite rod for AM.

At this point, that's about all the information on existing 4-way receiving radio systems currently available. Later in the year, Pioneer and other Japanese companies will possibly, come forward with more advanced models, although many are sure to follow the Motorola C-QUAM system for now rather than attempt either a Sony or Sansui multisystem IC.

Sony's Latest ICs

Just before publication we learned that Sony has now introduced two new multisystem AMstereo ICs. Identified as CX20177/CXA1017 automatic audio processors and the CX857 universal stereo decoder, they are offered for both auto receivers and stereo component systems. All are said to deliver fully automatic system selection, indication, and decoding for all four AM stereo systems. Either the CXA1017 or the CX20177 may be used, with CXA1017 being the smaller package that does *not* identify the particular stereo system being received. The CX20177 does. Both, however, identify AM stereo reception. Manufacturers prices range between $2.20 and $3.50 for quantity lots. The company made no further announcement of whether it will also offer complete mobile and home radio systems in addition to the two portables already on the market, but it said that more ICs for single standard receivers (probably C-QUAM, etc.) would be forthcoming later.

Chapter 6

Introduction to Multichannel TV Sound

It isn't difficult to become enthusiastic about television multichannel sound. After the advent of in-line picture tubes, small, efficient solid state chassis, comb filters and 4 MHz video passbands, the time has arrived for government and industry to authorize and engineer marked improvements in TV sound. Regardless of whether surveys have shown in the past that audience attention divides 70/30 between picture and music or speech, try sitting down with your favorite receiver and see how long you look at those fascinating camera revelations without a whisper spoken. You'll find that in less than a minute attention will wander and there'll be other things more attractive to do with your time.

Consequently, it's quite important to hear the spoken word, and considerably more so when there's quality music. The blending of separate reed and orchestra brass effects, for instance, can be thrilling for those whose ears have yet to be dulled by Naval guns or acid rock. And should this new medium produce excellent results, then many video/audio systems may be combined for maximum enjoyment and effects in forthcoming home entertainment and communications consoles.

What would you think, for instance, of a universal timer that turned television programs on and off, recorded sports, play and music events, and took care of home surveillance and police notification whenever you were away? That's just a sample of what's coming—with a great deal more in the works. Next to your automobile, the home entertainment center will eventually become your third greatest investment. With broadcast, cable, special services, and satellite inputs, it should pay off handsomely in both pleasure and profit. Audio, of course, assumes an increasingly important part.

HISTORY

The idea of stereo sound in television broadcasting is far from new. Prior to 1960, the National Stereo Radio Committee (NSRC) was formed by industry organizations to study the feasibility of AM, FM, and stereo TV broadcasting. Initially, of course, the main interest was in FM since its obvious base band and channel width offered a prime high fidelity transmission medium. So TV and AM proposals became dormant, and finally on April 19, 1961, the Federal Communications Commission adopted regulations for the FM Broadcast Service. The system selected was a combined product of Zenith and General Electric, with affected industry plants going into production shortly thereafter.

There has been interest in multilingual channels for Europe since the 1960s, and as far back as 1964, our own Federal Communications Commission issued a Notice of Inquiry in Docket No. 15697 for TV stereo. Television broadcasters didn't even reply, and there were few favorable comments, therefore that proceeding was terminated. Again, on July 1, 1977 the FCC reported a new Notice of Inquiry (Docket 21323), to which major networks ABC, CBS, and NBC replied there was insufficient public interest in stereo TV sound to justify such a service. But Public Broadcasting and TV station KERA reported enthusiastic public response to simulcasting TV and stereo FM, and there were "a number of inquiries from TV broadcast licenses concerning the possibility of using TV aural baseband subcarriers to provide specialized programming services to group with various interests." Telesonics, Inc., a TV stereo system developer, also stated that its system had been tested in Chicago on WTTW "with good results."

Americans then discovered that Japan had actually begun broadcasting stereo and bilingual programs during the later half of 1978. Following this, AT&T long lines offered one channel of 15 kHz high fidelity TV to the networks in January, 1978, with the promise that another one would follow. The Electronic Industries Association then began to awaken, and later in 1979 a subcommittee on multichannel TV formed to explore ways and means for the United States to take advantage of the medium. On June 30, 1981, the FCC adopted a First Report and Order (49 RR2d 1562) permitting TV aural baseband subcarriers for ENG (electronic news gathering) cuing and coordinating, with the exception that injection levels of any subcarriers could not exceed 15 percent, with system AM/FM noise levels limited to -50/-55 dB, respectively.

In this same report, the Commission "believes that it should now explore the possibility of further expanding the uses of the TV aural baseband"—with proposals allowing unrestricted operations "employing a wide range of technical systems," including multichannel TV sound for regular TV programming, program-related information for those with sight and hearing problems, "storecasting" and background music. But the FCC suggested that, initially, stereo might not be suitable for many TV programs, and, therefore, the stereo subchannel might be used for other purposes such as paging, electronic mail, facsimile office services, and municipal traffic light and sign control. Further, the Commission proposed that public broadcasters be allowed to offer "a full range of services on their aural baseband" since they have the same subcarriers as commercial broadcasters.

As for technical standards, the FCC believed that new subcarrier transmissions should offer a composite video/audio signal compatible with present mono receivers. Commissioners proposed aural carrier deviation of main channel should be maintained within 2 dB of the present level, a 60 dB attenuation of any subcarrier—other than the stereophonic subcarrier signal into the main channel—and crosstalk from the stereo subcarrier into the main channel should be -40 dB. They would also permit a pilot tone for stereo indication and switching, and would place an upper limit of 120 kHz on the aural baseband. FCC engineers did recognize that some of these suggestions and restrictions could pose technical difficulties for cable television because of greater bandwidths for multichannel sound transmission and the placing of several subchannels in the aural baseband spectrum—a supposition that proved correct, and a problem that has not yet been fully dealt with.

Zenith's View and Initial Tests

Sometime before the FCC's action, Zenith Electronics Corporation's Carl G. Eilers and Pieter Fockens made a significant proposal, taking into account past experience with FM radio stereo, SCA applications, subscription TV, and intercarrier buzz—of which there will be separate and detailed discussions once more emanating from the investigations of these two very able and experienced engineers. Fortunately, their Zenith multichannel proposal was fully accepted by the Electronics Industries Association (EIA), with some modifications, after intensive tests and over-the-air broadcasting via WTTW, Channel 11, Chicago in 1977. It has now been sanctioned by the Federal Communications Commission with stereo TV broadcasting just underway as the first copies of this book are published.

WTTW tests, however, were made with the pilot set at 1.25 f_h—with f_h being the TV horizontal scanning rate of 15,734 Hz/sec—the subchannel centered at 2.5 f_h (AM modulation), a separate audio program (SAP) FM modulated at 4.5 f_h, and ENG and TEL placed at 6 f_h and 6.5 f_h, respectively—both FM modulated. Even at this early stage, Larry Ocker of WTTW reported that "recent measurements . . . confirm that stereophonic TV sound of unprecedented quality can be implemented rapidly and cost-effectively using present day transmitters and current FM technology." Other participants were James Simanton and Carl Wegner of Telesonics Systems, Inc. During tests, it was found that video IF information contained AM and FM scan-frequency components (incidental phase modulation), resulting primarily from unequal sidebands in video transmitter and receiver front ends. A sound IF bandpass filter after the receiver tuner isolated the sound IF from video and "permanently" reduced sound channel noise level. In this way, both video and sound channels can be optimally tuned and controlled by AGC and AFC independent circuits for each channel.

Returning to Zenith, the initial proposition concluded that:

☐ Monophonic television sets should receive this stereo with no S/N loss and little reduction in video/audio quality.

☐ A second program "service" is required with stereo for bilingual or "other program related use."

☐ Stereo and monophonic audio should have equal fidelity (except for S/N), even though the "second program service" may have lessened quality.

☐ System parameters should look to the future rather than be shackled by the designs of today or the past.

A quick look at conventional FM radio shows a 19 kHz pilot signal to trigger stereo recognition in the receiver, L+R modulation of the main channel, and a 38 kHz suppressed carrier that's amplitude modulated by L−R. There's also often an SCA (storecast) background channel placed between 53 and 75 kHz, with an FM 67 kHz carrier. FM radio, of course, operates between 88 and 108 MHz, whether stereophonic or not, and has been allocated 200 channels by the FCC at a deviation of ±75 kHz per channel. Good antennas, apparently, do much to relieve the 23 dB S/N difference between FM mono and stereo. The general idea of FM radio stereo carries over into TV stereo sound, even including the AM modulated stereo subcarrier. But, because of television parameters, the two services are obviously different as you will soon see, even though each is capable of a full frequency response of 15 kHz.

Although "main and subchannel" provisions are similar to FM stereo, pilot and subcarrier frequencies are now defined as f_h and 2 f_h reserved for stereo only. A simultaneous second program channel could occur at 4 f_h, and nonpublic channels are also possible at 5.5 and 6.5 f_h. These subcarriers would be FM with low bandwidth audio or data signals such as frequency shift keying (FSK) or telemetry. The main channel in stereo would be the usual L+R, with the first subchannel becoming an AM double sideband suppressed carrier modulated by L−R.

The Zenith Proposal

Peak deviation of the main and stereo subchannel combined becomes 50 kHz, and the audio bandwidth would amount to 15 kHz with 75 microseconds pre-emphasis. With interleaving, the main channel peak deviation would also be 50 kHz so that mono sets retain the same level of loudness. The pilot could be sent at a 5 kHz main carrier deviation, while the second audio channel subcarrier appears at 5 f_h and is FM modulated to a peak deviation of 10 kHz, with the usual 75 μsec pre-emphasis. Subcarrier deviation of the main channel remains at 15 kHz, while first and second "professional" subcarriers have a 3 kHz main carrier deviation. Predicted distortion throughout the 15 kHz audio range would amount to less than 0.1 percent for the total system, made possible by an amplitude modulated subcarrier and not an FM subcarrier since with bandpass filtering, distortion would become greater than 3 percent even after "several kilohertz." Lesser quality services would appear, of course, in the FM carriers at multiples 4, 5.5, and 6.5 of the horizontal line rate. A diagram supplied shows all these carriers in terms of aural carrier frequency deviation which is graphed between 0 and 50 kHz. Note that ENG and Telemetry are placed down the deviation response (Fig. 6-1). Some of these terms, including peak deviation and pilot frequency, were doubled from the original Zenith proposal prior to EIA testing.

The EIA (of) Japan Proposal

Japan's EIA was also interested in having its system examined and adopted by the U.S. EIA Broadcast Television Systems Committee (BTSC) which began work in January 1979. A proposal was therefore submitted to the Multichannal Sound Subcommittee of BTS and system examinations got underway. Japan of course, was already on the air at home and had the advantage of some experience with multisound.

The Japanese system is somewhat similar to that proposed by Zenith, but uses FM rather than AM modulation for the subcarrier at 2 f_h, with h being the same 15,734 Hz line scanning frequency as before. The first subchannel would be modulated by L−R to a peak deviation of 10 kHz and lies in the frequency band between 16 and 47 kHz. A delay of

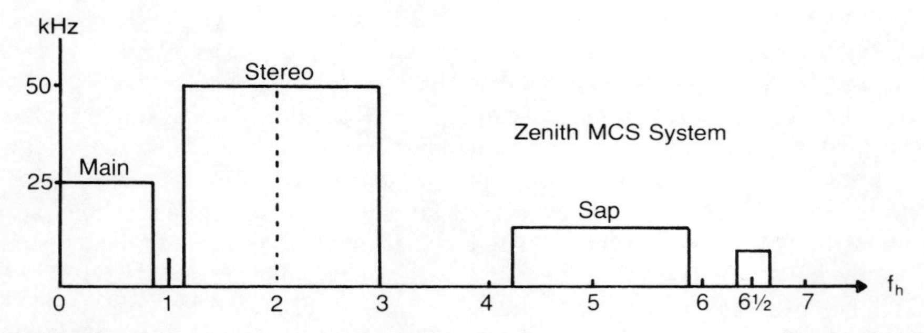

Signal Specifications							
Service or Signal	Modulating Signal	Max. Mod Freq. kHz	Pre-Emphasis μsec	Sub-carrier Freq. *kHz	Type of Modulation of Subcarrier	Subcarrier Deviation kHz	Main Carrier Peak Deviation kHz
Monophonic	L+R	15	75				25
Pilot				f_h			5
Stereophonic	L−R	15		2 f_h	AM-DSB-SC		50
2nd Program		10		5 f_h	FM	10	15
Prof. Ch.	Voice or	3.4	150	~6½ f_h	FM	3	3
	Data	1.5	0	~6½ f_h	FSK	3	3
						Total	101

*f_h = 15.734 kHz

Fig. 6-1. Block and signal specifications of the initial Zenith stereo TV sound system proposal (courtesy Zenith Electronics Corp.).

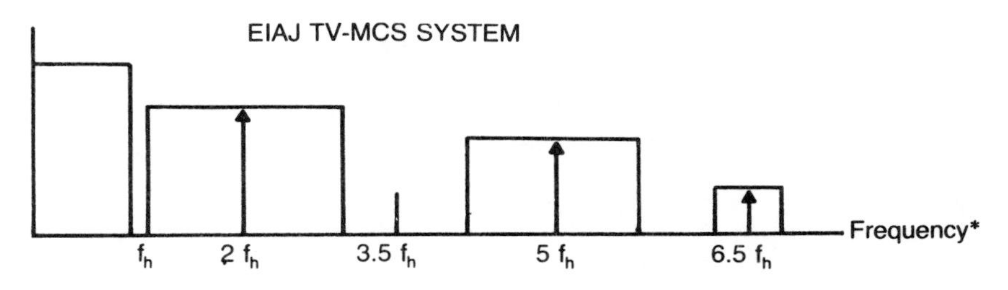

Signal Specifications							
Service or Signal	Modulating Signal	Max. Mod. Freq. kHz	Pre-Emphasis μsec	Subcarrier Freq. kHz*	Type of Modulation of Subcarrier	Subcarrier Deviation kHz	Main Carrier Peak Deviation kHz
Stereo							
Main Ch.	L+R	15	75	2 f_h			25
1st Sub-Ch.	L−R	15	75	3.5 f_h	FM	10	20
Pilot							2
SAP 2nd Sub-ch.	2nd Program	10	75	5 f_h	FM	10	15
Prof. Ch.	Voice	3.4	150	6.5 f_h	FM	3	3
	or Data	2	0	6.5 f_h	FSK	3	3
						Total	65

* f_h = 15.734 kHz

Fig. 6-2. Initial Japanese block and signal diagram using FM carriers (courtesy EIA Japan).

20 μsec compensates the main channel because of subchannel decoder delay. Main and subchannels are deviated 25 and 20 kHz, respectively, and a 3.5 f_h CW pilot signal deviates the main channel with 2 kHz to identify stereo (Fig. 6-2).

The second subchannel has a subcarrier at 5 f_h with peak deviation of 10 kHz and a 10 kHz audio bandwidth, occupying the frequency band between 68 and 89 kHz, modulating the main carrier to 15 kHz deviation. Located at 6.5 f_h, the professional channel deviates the main carrier to 3 kHz with a pre-emphasis of 150 μsec versus 75 μsec for other channels. At a modulation frequency of 2 kHz with zero pre-emphasis, the professional channel could carry FSK with subcarrier and main carrier peak deviation of 3 kHz. All signals combined would result in a total main carrier deviation of 55 kHz as illustrated in Fig. 6-2.

The Telesonics Proposal

L+R would FM modulate the main channel with left and right inputs and an unmodulated pilot carrier at 5/4 f_h (19.67 kHz) would peak-deviate the main carrier 0.8 kHz. An amplitude modulated suppressed subcarrier at 5/2 f_h (39.335 kHz), phased to the pilot signal, would be amplitude modulated by L−R inputs generating double sidebands—with both modulations pre-emphasized 75 μsec at a minimum audio bandwidth of 15 kHz.

While the peak main carrier deviation amounts to 50 kHz, main and subcarrier deviations would amount to 50 kHz also; because of interleaving. The second subchannel would be frequency modulated with subcarrier at 9/2 f_h, or 70.8 kHz and an audio bandwidth of 3.4 kHz, peak-deviation the main carrier at 3 kHz. Pre-emphasis for this subchannel would be 150 μsec whether ENG or telemetry, but telemetry would have a 2 kHz maximum modulating frequency and a subcarrier and main carrier deviation of 2 kHz. The third subchannel, also frequency modulated, could carry audio, have a maximum modulating frequency of 10 kHz, pre-emphasis of 75

μsec, a subcarrier frequency of 6 f_h, and a peak carrier deviation of 10 kHz and deviate the main carrier by 15 kHz. Figure 6-3 amply illustrates all channel positions, their operating frequencies, and other parameters.

The foregoing pretty well wraps up some of the historical portions and beginnings of the program which was a very deliberative action on the

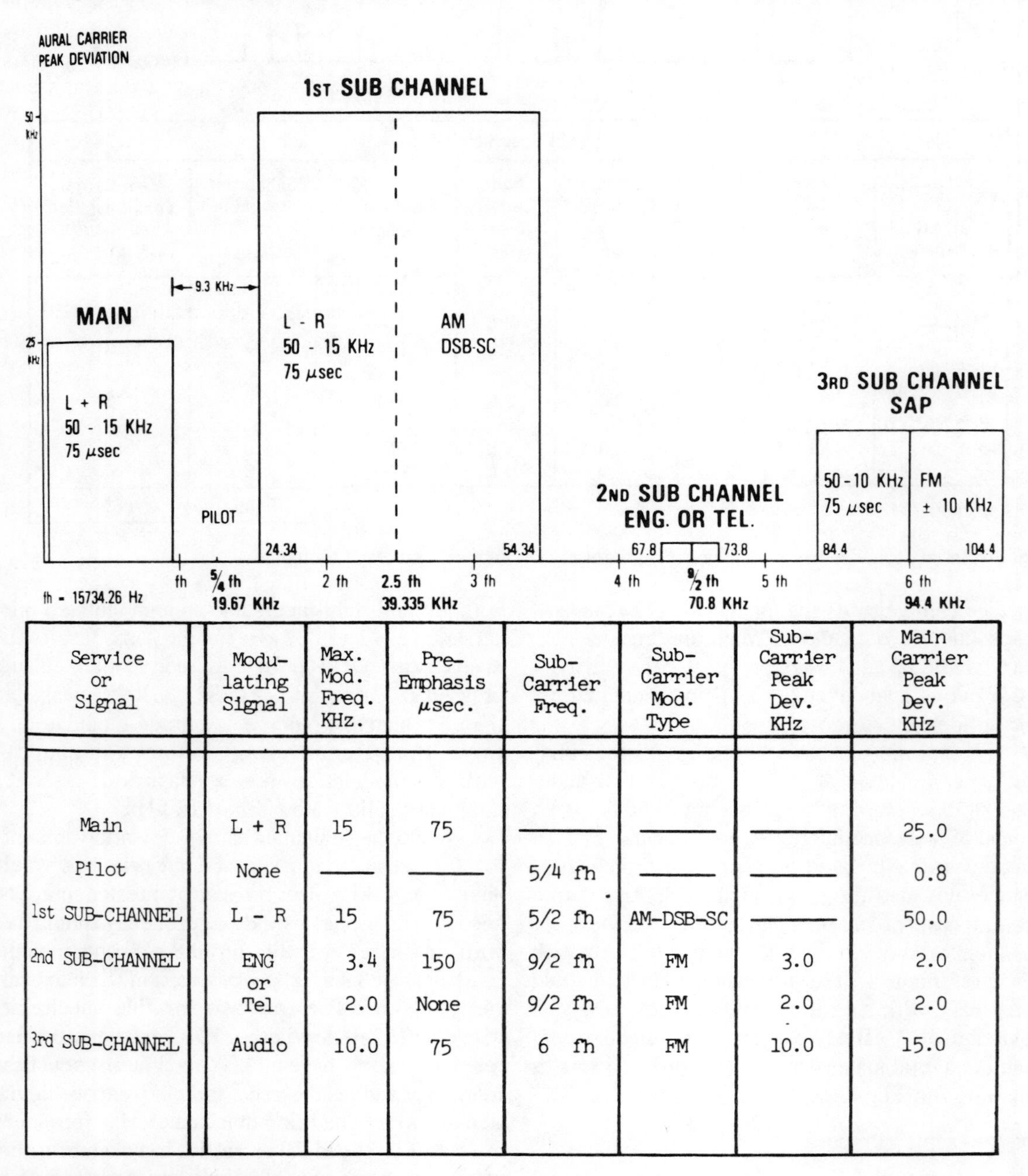

Service or Signal	Modulating Signal	Max. Mod. Freq. KHz.	Pre-Emphasis μsec.	Sub-Carrier Freq.	Sub-Carrier Mod. Type	Sub-Carrier Peak Dev. KHz	Main Carrier Peak Dev. KHz
Main	L + R	15	75	—	—	—	25.0
Pilot	None	—	—	5/4 fh	—	—	0.8
1st SUB-CHANNEL	L - R	15	75	5/2 fh	AM-DSB-SC	—	50.0
2nd SUB-CHANNEL	ENG	3.4	150	9/2 fh	FM	3.0	2.0
	or Tel	2.0	None	9/2 fh	FM	2.0	2.0
3rd SUB-CHANNEL	Audio	10.0	75	6 fh	FM	10.0	15.0

Fig. 6-3. Telesonics multichannel sound system.

part of the Electronic Industries Association, its engineering members, and the three proposers. In the meantime, the FCC observed most or all proceedings but was generally a bystander in analyzing and testing of the three proposed systems.

COMPANDING

We've covered a general outline of system specifications and their objectives—along with a few impediments. Compression and expansion (companding) for noise reduction is used to escape "sub-standard stereo service to these listeners beyond the Grade A contour." As late as the summer of 1982, the EIA/BTS sound group demonstrated to a "number of TV manufacturers" a stereo television sound simulation without noise rejection. All agreed that anyone living in a Grade B area would have to use monophonic for acceptable sound quality. Therefore it was decided that some type of dynamic range system would be required for stereo TV sound where 47, 56, and 64 dBu Grade B service was not acceptable, and which would also serve Grade A designations of 68, 71, and 74 dBu as well.

So, to improve the dynamic range of the stereo TV system, a means had to be found to handle both loud and soft sounds at both transmitter and receiver. Such a range can be the difference between overload (when 3 percent harmonic distortion appears) and a weighted noise floor. EIA reported that *without* companding, the dynamic range of stereo TV "varies directly with the signal strength" at the receiver, limited only by thermal noise and FCC imposed signal limits.

A compander must then make the ear believe there's extended dynamic range when there isn't and also pass quality audio waveforms throughout the compression/expansion cycle without "significant degradation." All this, obviously, is somewhat difficult to do and much depends on design and the amount and kind of interference present . . . the final objective being to process audio programs faithfully and with minimum background noise and/or interference. Naturally, there are many ways to do this, and the EIA had to examine several systems, among them, CX-TV, dbx, and Dolby. We'll look at these in the order given, paraphrasing the EIA report.

The CX-TV System

This system consists of a "professional" compressor and expander, and a consumer-type expander with a dedicated integrated circuit. It uses wideband amplitude compression and pre-emphasis as well as wideband expansion and de-emphasis. Expansion threshold is 10 dB below 100 percent modulation, offering immunity to "extraneous" noise while signals above -10 dB are expanded by 1:3. Total noise reduction is greater than 30 dB for both stereo difference and separate audio program channel (SAP). The same expander may be used for both stereo and SAP.

Background noise reduction amounts to 20 dB, with another 12 dB added with pre-emphasis and de-emphasis. Signals greater than -30 dB are compressed with respect to their reference and above 500 Hz with a special signal-dependent variable time constant.

Expansion occurs at 1:3 for signals above -10 dB with respect to reference and above 500 Hz, with the same signal-dependent time constant, offering good noise masking. As you can see in Fig. 6-4, compression and expansion circuits are exactly the inverse of one another using the same control circuits. Pre-emphasis and de-emphasis times are given as 320 μsecs.

For low cost and simplicity in television receivers, a dedicated IC Hitachi HA12044 expander and 14 external components are available and said to perform creditably.

The dbx, Inc. System

Once again, complementary encoder/decoder functions makes *dbx* a look-alike for transmit-receive operations and proponents claim it can deal flexibly with many audio problems such as hum, buzz, whistles and hiss. Compression and expansion (compansion) is done on a 2:1:2 basis, keeping signal levels high during transmission with pre-emphasis "hinged" about 1 kHz and a range at high frequencies of ±25 dB, accompanied by peak slopes at 12 dB/octave. It's said to have "the widest range of independent gain/spectral tilt combinations of any of the proposed NR (noise reduction) systems

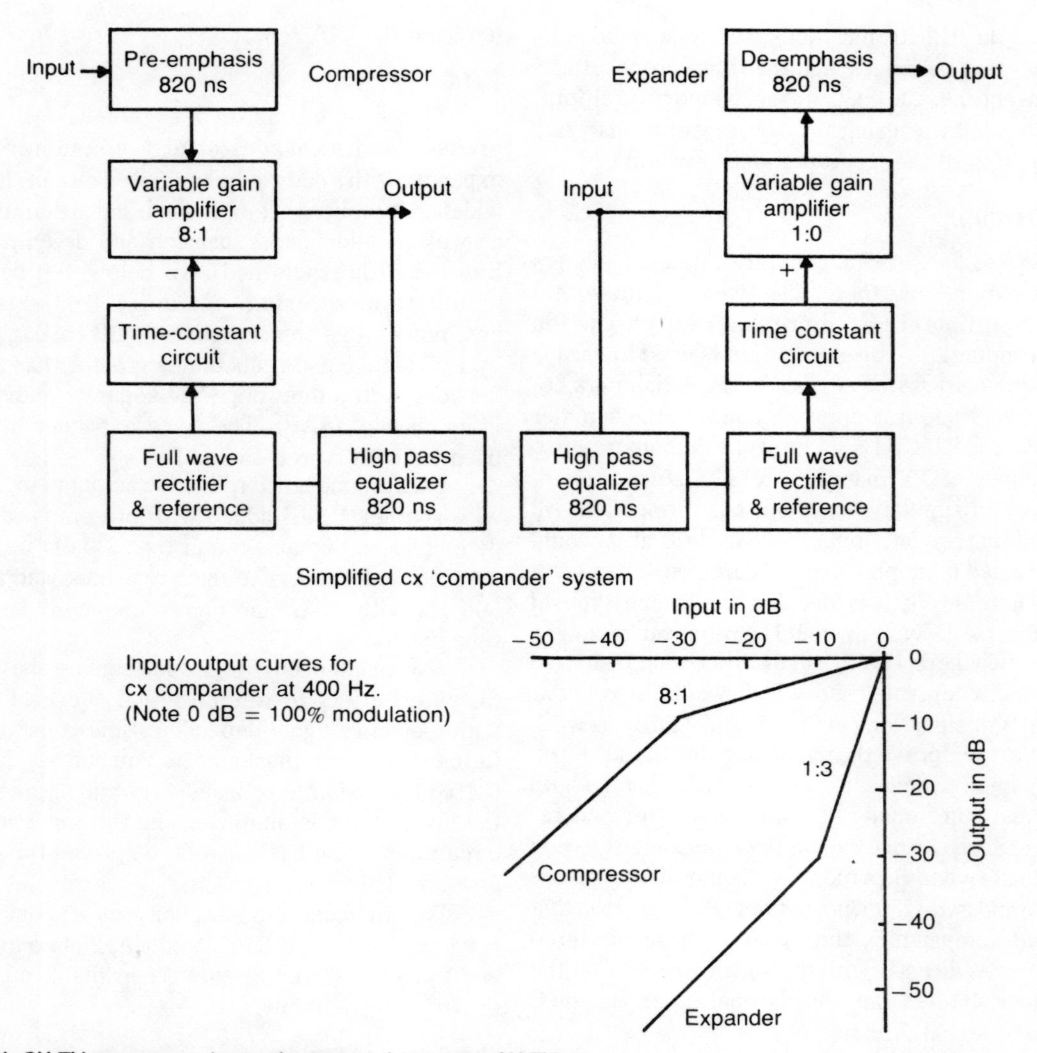

Fig. 6-4. CX-TV compressor/expander proposal (courtesy CX-TV).

and can deal with the widest range of signal/noise spectral combinations without overload or overmodulation." Figures 6-5A and 6-5B illustrate both compressor and expander as drawn by dbx, Inc.

The encoder lowpass filters remove extraneous signal material. With wideband compression it keeps signal levels above transmission noise and avoids overmodulation, compresses spectral variations performs static pre-emphasis, and eliminates peak overshoots by pre-emphasis clipping.

The Encoder. As shown in Fig. 6-5A, the encoder supplies a 2:1 compression range at low and mid-audio frequencies where "channel dynamic range" is greatest and most signal energy collects. But at high frequencies with constricted dynamic range, the compression ratio becomes 3:1, with more rapid time constants.

The two semi-independent (same source) root-mean-square (rms) detectors sample signal levels and high frequencies, adjusting wideband compression and dynamic pre-emphasis through the variable gain element and pre-emphasis circuits. These rms detectors permit program-dependent time constants to handle audio variations over the complete compression/expansion cycle.

The wideband compressor signal is band-limited to between 100 Hz and 3 kHz, while dynamic pre-emphasis operates on signals from 4-9 kHz. Low end rolloff for this control signal is especially sharp so it will respond to high frequency information even with strong low and middle frequency incoming signals. The clipper eliminates large, short-duration transients in the compressed and pre-emphasized intelligence. Since it is within the feedback loop and located before the rms detectors, FCC modulation requirements are compiled with and there's no compromise in sound.

The Decoder. The simplified block diagram

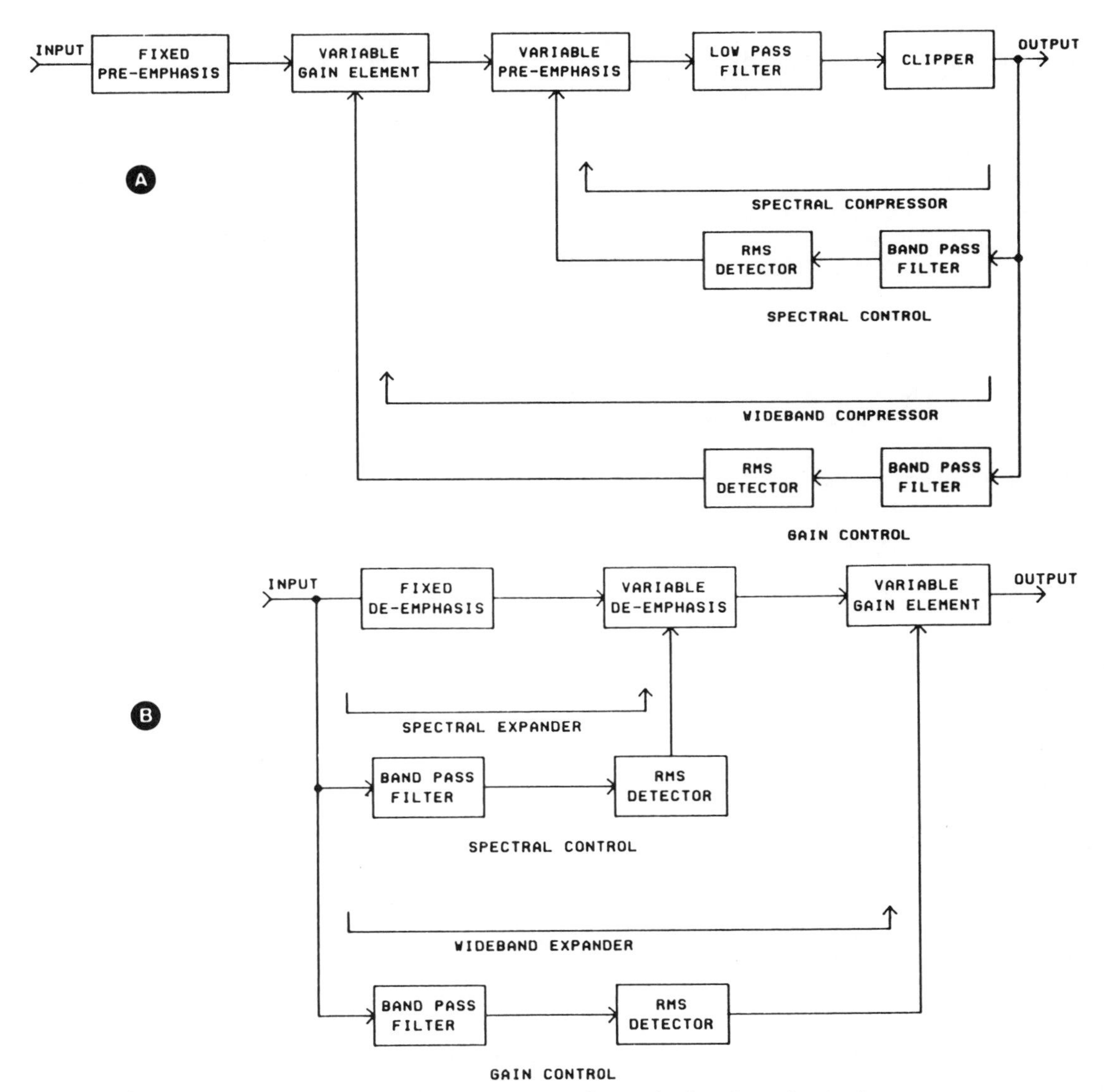

Fig. 6-5 A and B. Composite block diagram of dbx compressor/expander (courtesy dbx, Inc.)

of this circuit is illustrated in Fig. 6-5B. According to dbx, the encoder is fully complemented and both spectral and amplitude dynamics are restored to the initially modulated signal. There are wideband spectral/amplitude expansion and static de-emphasis of 1:2 at low and mid frequencies, increasing to 3:1 at high frequencies. Rms levelers in the expander monitor, process the incoming compressed information and never track anything but the compressed signal. There is no "unlimiting," some 8 dB of "headroom" if permitted for static signals, and clipping only operates on small-energy transients with little signal attenuation. As you can see, the expander looks and operates much like the compressor over its entire dynamic range.

In describing dbx, we occupied more space than with its competitors since this is the system chosen to operate with Zenith's transmitter/receiver, and it deserves a little more attention.

The Dolby System

This is a triple "path" system offering unity gain in one path and two parallel paths controlling individual frequency spreads. Compression occurs when all three paths are summed, while expansion takes place when the parallel paths are situated "in a self-correcting negative feedback loop." Both encoder and decoder operations are outlined in Fig. 6-6.

Dolby Laboratories explains this relationship in terms of unity gain and side-chain paths as shown in the illustration. Side-chain 1 appears as a fixed band limiter having a fundamental bandpass range between 40 and 600 Hz, followed by a voltage-controlled attenuator (vca) providing gain reduction. An amplifier then routes this signal to a nonlinear limiter as well as a full wave rectifier, the latter's output being integrated via a nonlinear two-pole filter and returned as control to the voltage-controlled attenuator. A second input to the full wave detector also reaches the rectifier from the bandpass filter in a feed-forward path where the two inputs are probably level detected to establish outputs for various gains.

With low signal inputs, the voltage-controlled attenuator is threshold limited to a fixed minimum level. Side-chain 1 furnishes maximum gain of 18 dB which contributes mainly to compression output. But as compression inputs increase, vca attenuation also increases until side-chain gain is less than unity. At higher signal inputs, noise reduction should only become a small proportion of the unity-gain path; therefore some of the input signal reaches the full wave rectifier. More will do so with additional signal strength, "producing a downturning curve in the side-chain transfer characteristic." Consequently, at very high levels, the unity-gain path is the "dominant contributor to overall compressor output."

Side-chain 2 is described as a frequency-selective limiter with sliding-band filter characteristics. Its first order high pass response has a cutoff of 600 Hz, and once again there's a full wave rectifier and nonlinear filter. A following amplifier adjusts limiter contribution to the compressor output, driving both pre-emphasis and a nonlinear limiter before summation. Once more, filter control comes from integrated rectifier output through the two-pole, nonlinear filter.

When compressor inputs are below threshold, side chain 2 response is that of a 600 Hz high pass filter with 18 dB maximum gain, and side-chain signal dominates compressor outputs above 600 Hz. But as compressor levels increase cutoff, frequency levels increase, and there's a reduction in side-chain gain so that at high levels side-chain 2 contributes little to the compressor output. When there is actual program material, a 75 μsec pre-emphasis network enters into the control path. Overshoots are controlled by a nonlinear integrator, while a nonlinear limiter prevents transients overshoots of noise reductions during compressor attacks.

The expander receives the three summed inputs and goes through the compression mode in reverse. Sliding and fixed bands are apparent, as is the unity gain path and the two side-chain outputs. Identical nonlinear limiters appear also in the expander and their products are subtracted from expander signals and appear as precise replicas of the compressor inputs. De-emphasis, of course, is also 75 microseconds.

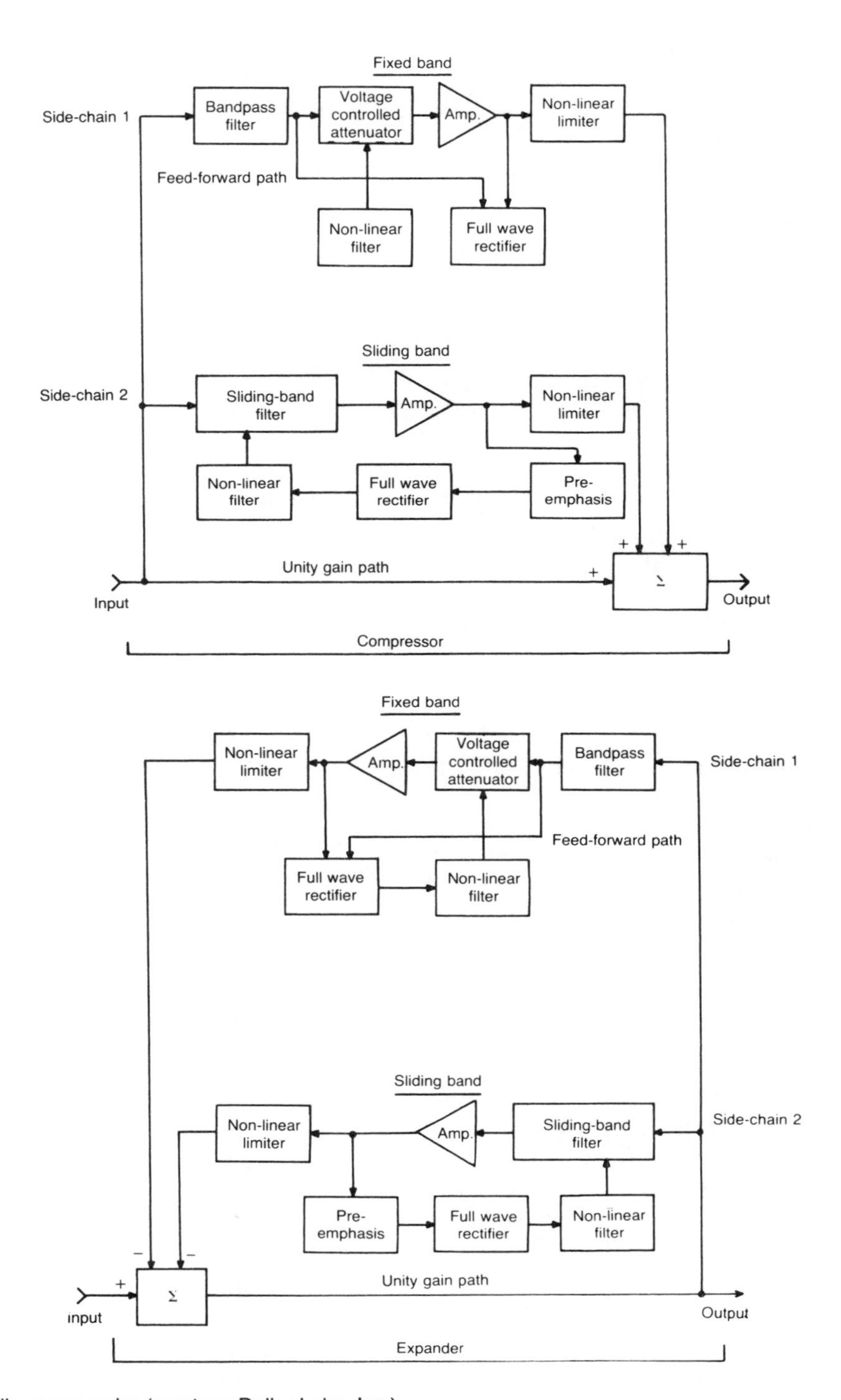

Fig. 6-6. The Dolby compander (courtesy Dolby Labs, Inc.).

System Commonalities and Differences

For extra edification we'll list at least most of the common parts of these transmissions and their differences as viewed by the industry before final testing and ultimate EIA and FCC selections. The chapter then concludes with another concise look at somewhat typical encoder and decoder, even though in mid 1984, final versions had not been released for publication. Remember, the systems discussed so far in this introductory chapter have been preliminary types used for engineering and evaluation and decidedly do *not* necessarily represent the electronics you'll see and hear in following years. In fact, when digital television arrives, you may see substantially different electromagnetic signals over the air and very different receivers to decode and display them.

So after those words of prudent caution, let's list the commonalities in reasonable order, followed by some obvious dissimilarities:

Commonalities. The monophonic audio signal receives an extra supersonic carrier for FM multiplex on the main carrier, in addition to multiples of this carrier for subcarriers of stereo difference signals for a second audio program (SAP), and a third for Telemetry of "professional" voice-grade transmissions. For compatibility, monophonic standards are unchanged, including power and peak deviation, but companding is proposed in L−R and SAP channels for background noise improvement. Usually, the initial supersonic carrier frequency selected is the same as television's present horizontal scan frequency of 15,734 Hz.

Differences. Japan's EIAJ FM modulates the stereo difference signal while U.S. manufacturers carry L−R on amplitude modulation with suppressed carrier. Other carriers in the U.S. systems are traditional FM. There are frequency injection and deviation differences between the three systems also, with Zenith-Telesonics having a combined mono and subchannel injection of 50 kHz. As you have seen, the three companding systems are quite different in engineering details as well as results.

TEST ENCODERS AND DECODERS

The following block diagrams and very brief descriptions characterize the various encoder and decoder systems used by the several proponents in their quest for EIA/FCC approval of multichannel sound. There will, of course, be some repetition of recent paragraphs describing these same systems, but additional block diagrams illustrate in more detail what these electronics do and how they do it. It's something that students or working engineers may require in examining early attempts to evolve satisfactory TV stereo sound systems and avoid problems the pioneers didn't foresee.

Also it's well to remember that a long time ago somebody *else* invented the wheel, the flatbed, and an engine to pull it. What's needed now are improvements or entirely new systems to do a better job during the ongoing communications explosions in *all* disciplines. Stereo sound for television is no more and no less than a rapidly developing part of the whole. Taken together, communications pictures become vivid, useful and complete. The next TV step should be an almost doubling of luminance and chroma bandwidths—then you'll both hear and see remarkable sound and detail. By 1987 all this should come to pass.

EIAJ Encoder/Decoder

The encoder shown in Fig. 6-7A begins left and right inputs with 75 microsecond pre-emphasis, is matrixed or not matrixed for mono or stereo, passing through limiting amplifiers and a time delay circuit before final 15 kHz low pass filters. L−R stereo is then 2 × 15.734 kHz FM modulated and passes through a bandpass filter circuit of 16 to 47 kHz before reaching the adder and baseband signal output. L+R, of course, is the main channel, operating in either mono or stereo modes. A crystal-controlled 3.5 f_h (horizontal line frequency) becomes the pilot for stereo, and SAP (second audio) FM modulates the second subcarrier at 5 f_h. Telemetry and/or ENG modulates the third subcarrier at 6.5 f_h which operates only on low frequency or voice and data information. In the latter instance, normal 75 μsec pre-emphasis is doubled, as all signals pass through appropriate low pass filters and bandpass filters.

Sync separation and automatic phase control

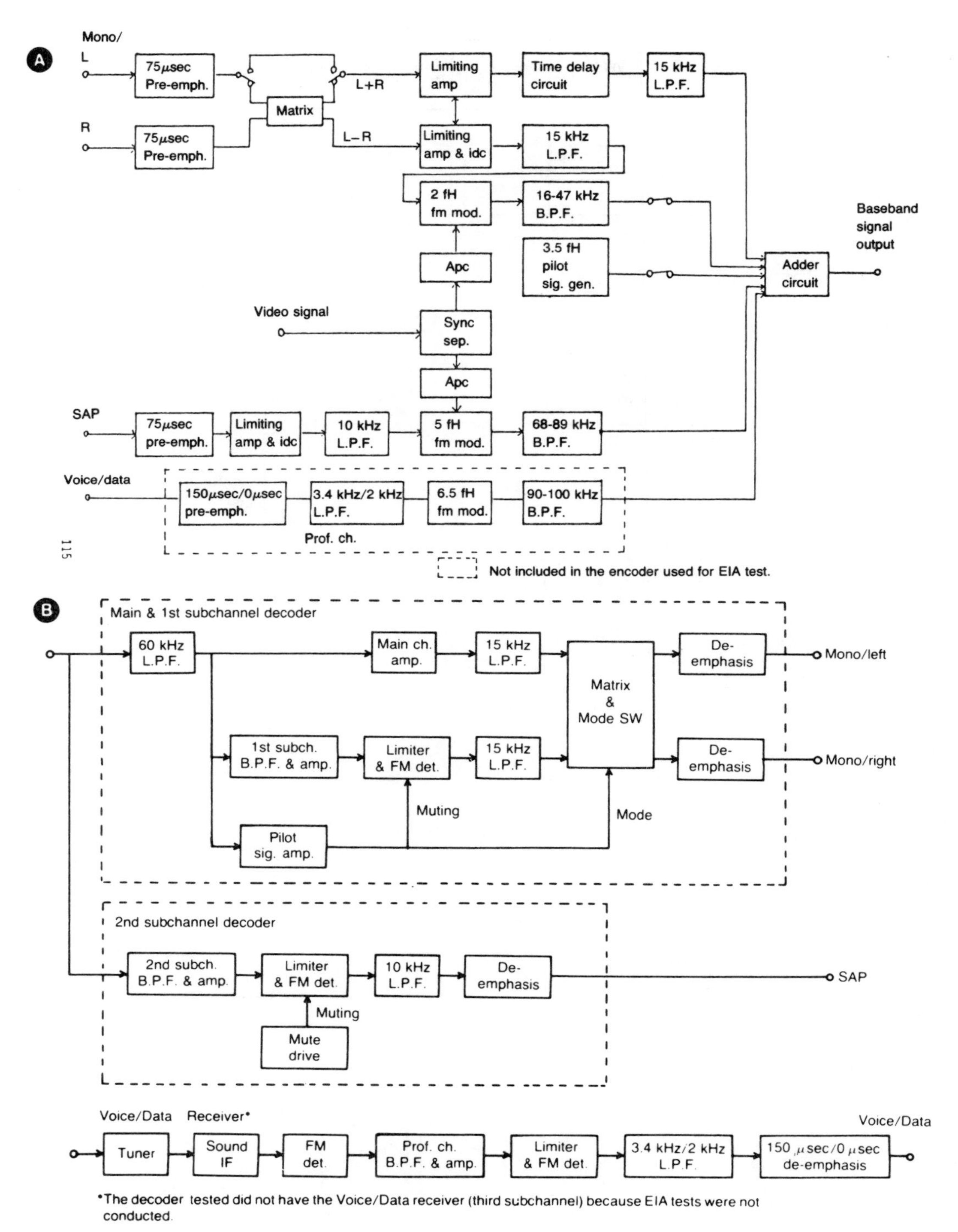

Fig. 6-7 A and B. The EIAJ encoder/decoder expanded signal diagram (courtesy EIAJ).

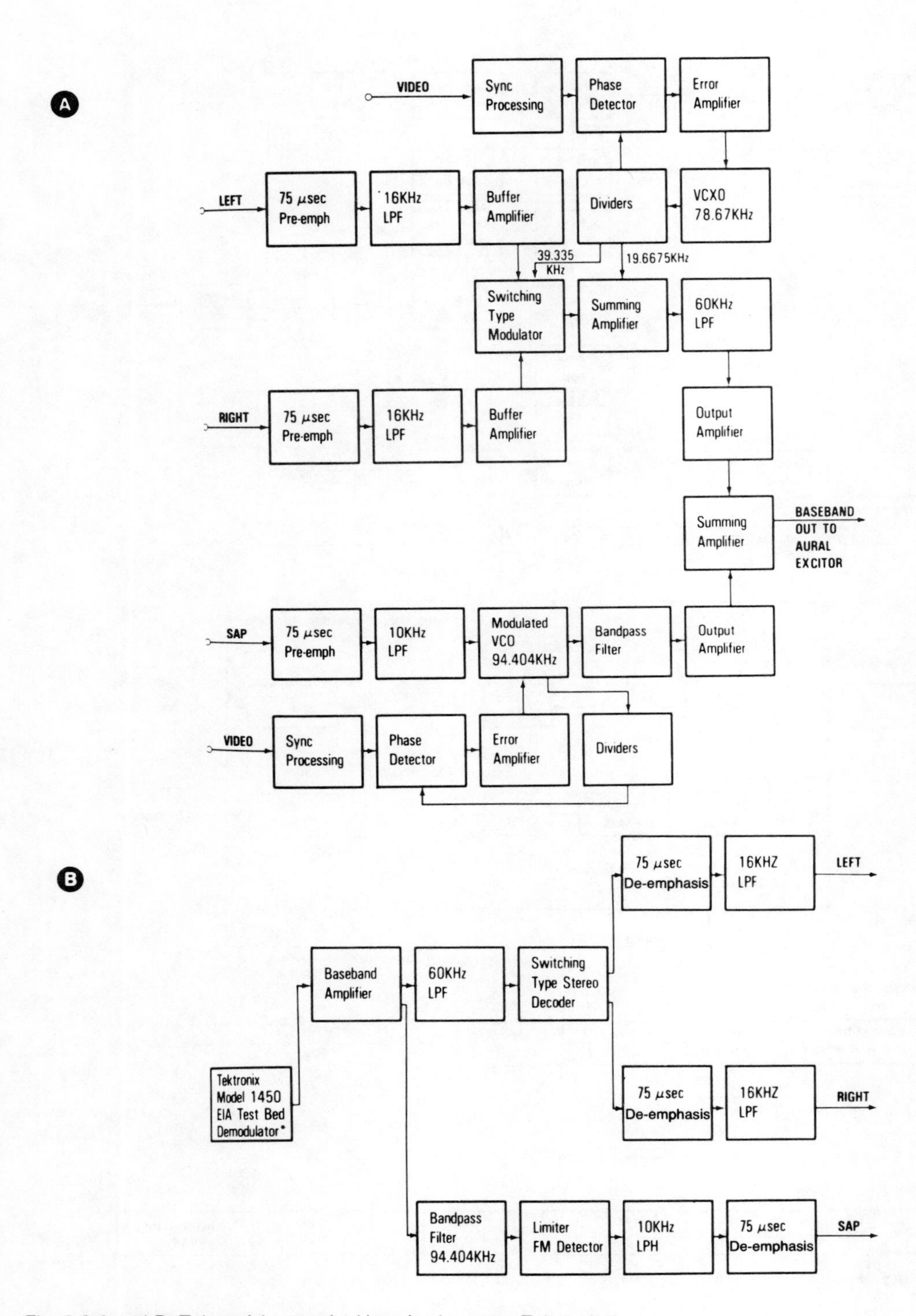

Fig. 6-8 A and B. Telesonic's encoder/decoder (courtesy Telesonics).

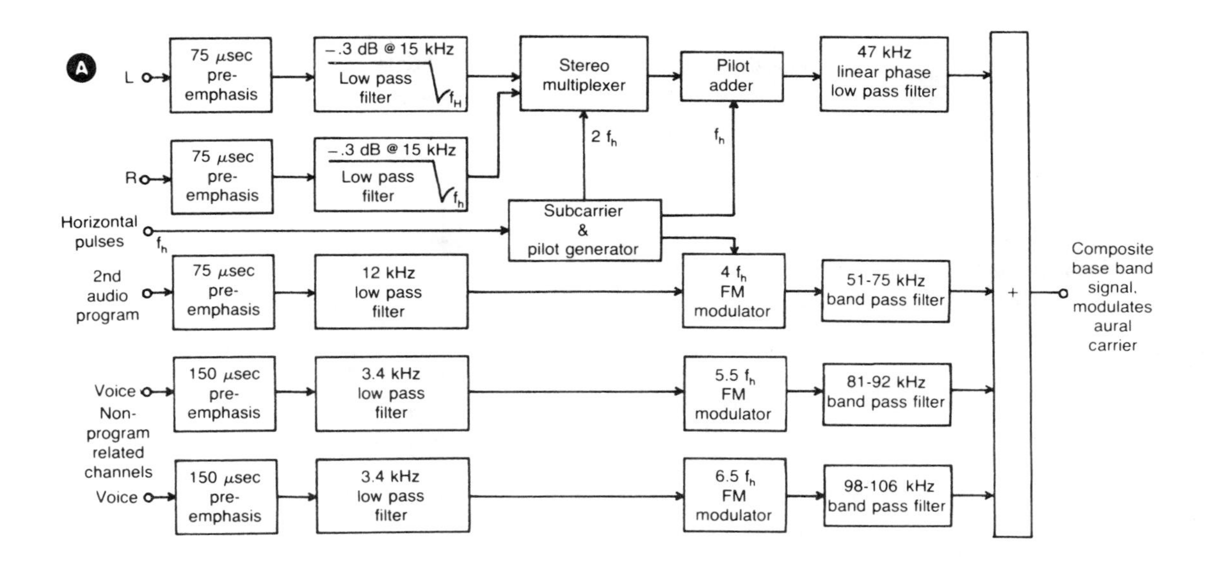

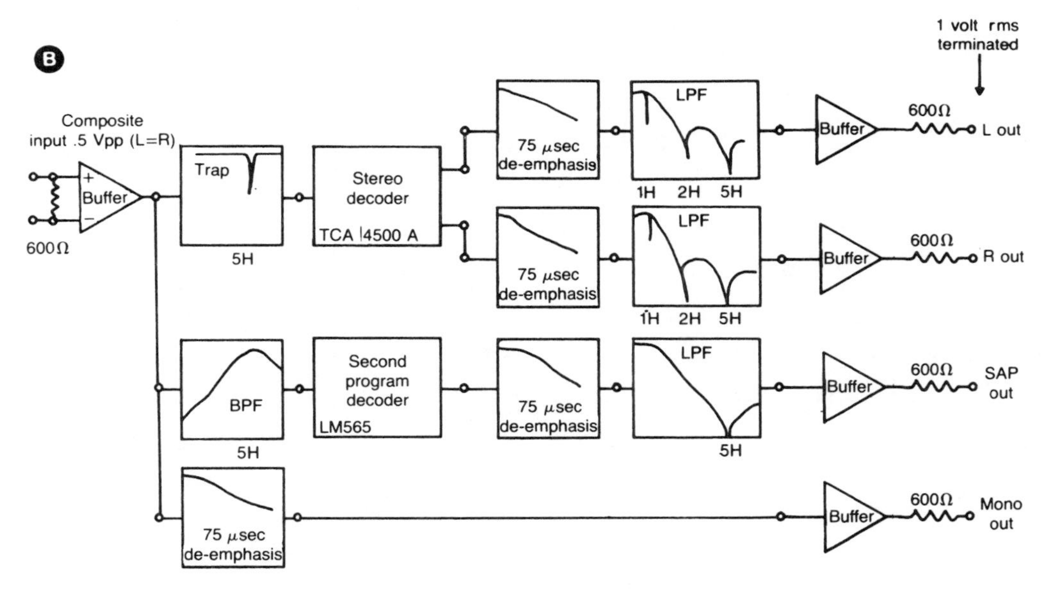

Fig. 6-9 A and B. Combined Zenith encoder/decoder block diagrams with various signal paths (courtesy Zenith Electronics Corp.).

are derived from the incoming composite video signal which contains transmitted sync pulse at both vertical and horizontal rates during their respective 1.4-millisecond and 11-microsecond blanking intervals.

The decoder (Fig. 6-7B) receives incoming information into a 60 kHz low pass filter for the first subchannel and main channels and also through a bandpass filter and amplifier for the second subchannel. Although the voice/data receiver is shown with tuner, IF, FM detector, plus professional channel bandpass amplifier and limiter/detector,

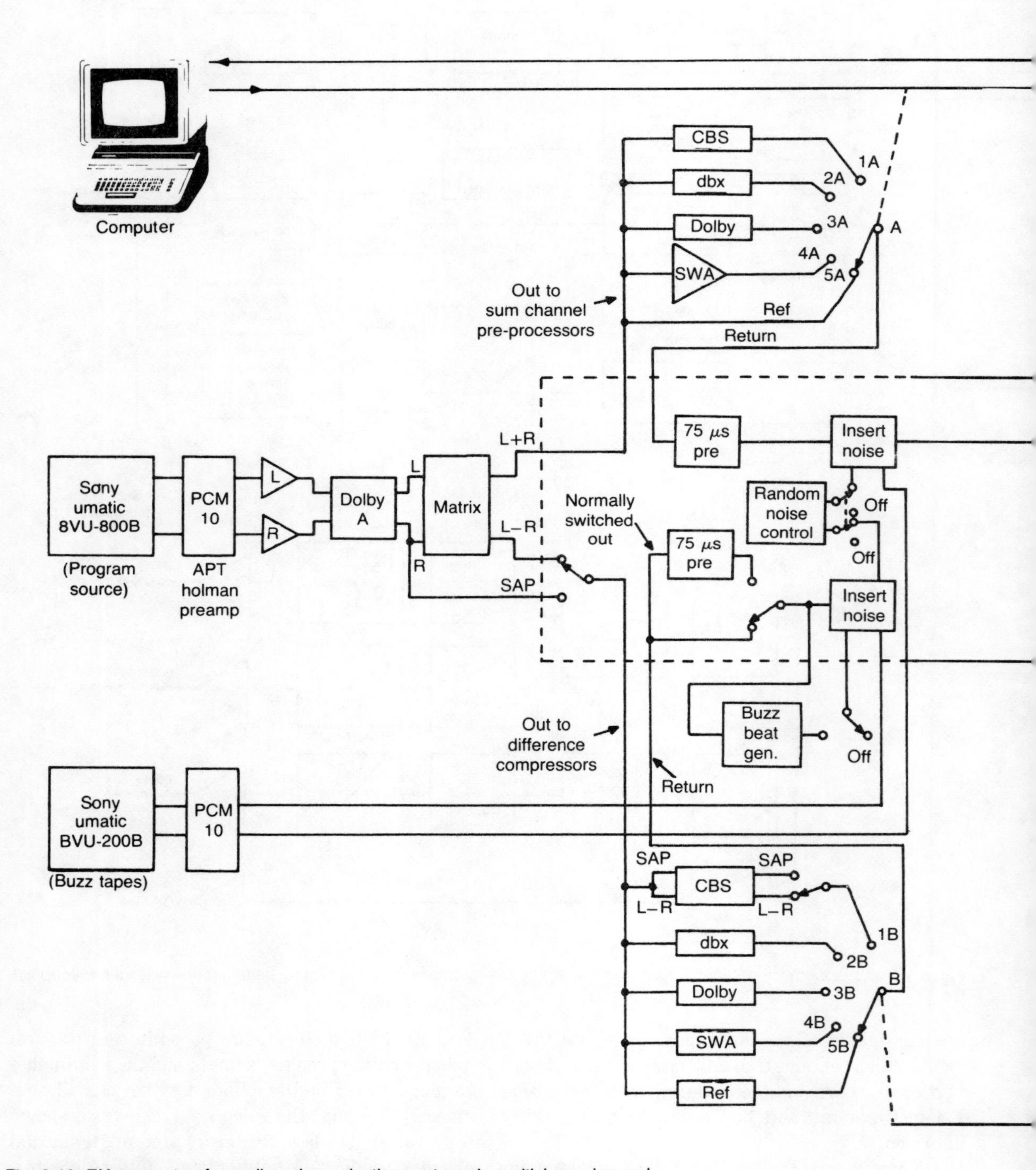

Fig. 6-10. EIA test setup for audio noise reduction systems in multichannel sound.

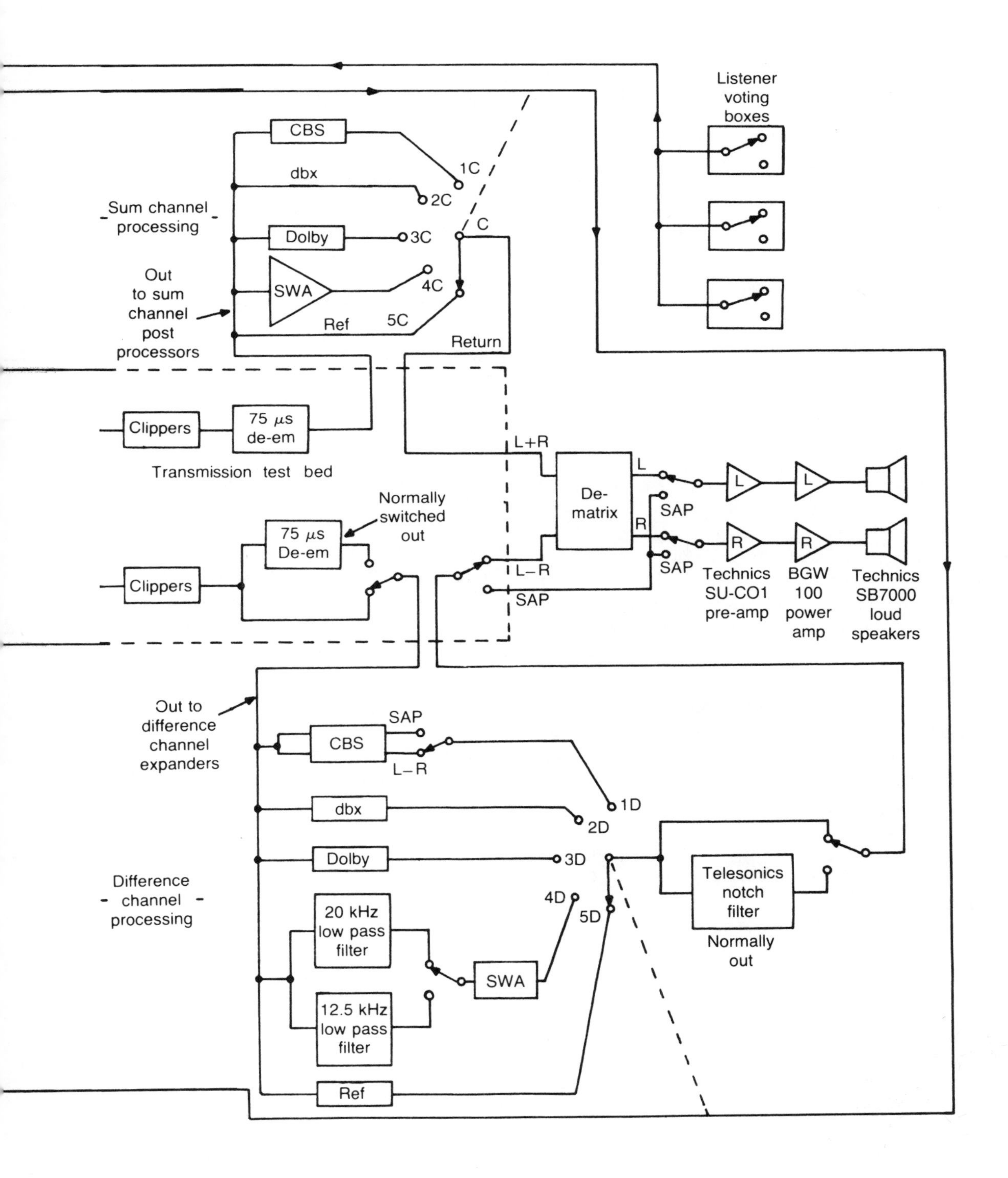
Listener
voting
boxes
CBS
dbx
1C
2C
Sum channel
processing
Dolby
3C
C
SWA
4C
Out
to sum
channel
post
processors
Ref
5C
Return
Clippers
75 µs
de-em
Transmission test bed
L+R
De-
matrix
L
R
SAP
SAP
L
L
R
R
Normally
switched
out
75 µs
De-em
Clippers
L−R
SAP
Technics
SU-CO1
pre-amp
BGW
100
power
amp
Technics
SB7000
loud
speakers
Out to
difference
channel
expanders
SAP
CBS
L−R
1D
dbx
2D
Dolby
3D
Telesonics
notch
filter
Normally
out
Difference
channel
processing
4D
5D
20 kHz
low pass
filter
SWA
12.5 kHz
low pass
filter
Ref

the decoder tested did not have this receiver and therefore EIA did not test this portion.

First and main subchannels are split after the 60 kHz filter, and the pilot signal and 1st subchannel are removed for processing in the limiter and FM detectors and Matrix/Mode switch. After 75 μsec de-emphasis, the right and left or monophonic components are extracted. In the second subchannel, incoming information passes through a bandpass filter, is limited and FM detected (muted, if necessary) and then limited and de-emphasized for separate audio program channel (SAP).

Telesonics Encoder/Decoder

Telesonics places video inputs top and bottom (Fig. 6-8A) for sync processing, phase detection, and error amplification, then introduces right, left and SAP inputs independently through 75 μsec pre-emphasis networks. Sync inputs maintain the 78 and 94 kHz of the two voltage controlled oscillators as right/left signals pass through 16 kHz low pass filters, and the separate audio is routed through a 10 kHz LPF. Buffer amplifiers deliver L/R signals into the switch-type modulator operating at 39.335 kHz, as this frequency is halved for the summing amplifier for passage through the 60 kHz low pass filter, output amplifier and into the summing amplifier. The summing amplifier also receives SAP information modulated on the 94.404 kHz vco, which is bandpass filtered before summing. Dividers from the vco keep the phase detector on track as it monitors the vco and transmits an error signal if phase or frequency varies.

In receive, a Tektronix 1450 demodulator supplies baseband-amplified signals into 60 kHz and 94.4 kHz filters, where left and right signals are decoded and the SAP information limited and demodulated by an FM detector. All outputs are then low pass filtered and 75 μsec de-emphasized before final audio amplification. Telesonics says the baseband amplifier would not be required for the usual consumer products. See Fig. 6-8B.

Zenith Encoder/Decoder

Zenith, too, pre-emphasizes its L, R, and SAP channels at 75 microseconds and the two voice or data channels at 150 μsec. See Figs. 6-9A and 6-9B. After the 15 kHz low pass filters, left and right audio portions are stereo multiplexed along with 2 f_h modulation from a horizontal pulse-driven pilot signal and subcarrier generator which also serves the pilot added at f_h and the SAP program deviating the 4 f_h FM modulator. Both voice and data channels are routed through 3.4 kHz low pass filters and on to 5.5 and 6.5 f_h modulators. Before summation in composite baseband, these four respective modulations are 47 kHz low pass filtered, and 51-75, 81-92, and 98-106 kHz bandpass filtered.

Decoding begins with an amplifier-buffer input proceeding to a 5 f_h trap, a 5 f_h bandpass filter, and the usual 75 μsec de-emphasis circuit and buffered monophonic output. This is all shown in Fig. 6-9B, as stereo and SAP program material pass through their respective decoders, are de-emphasized at 75 μsec, and then routed through three low pass filters and their buffered 600-ohm outputs. Note de-emphasis slopes and low pass filter rolloff curves and their f_h, 2 f_h and 5 f_h notches.

We won't belabor the testing procedure other than to say that the Multichannel Sound Committee did specify and construct a well-defined "test bed" for their investigations at the Matsushita Industrial Co. in Chicago (Fig. 6-10) to interface with proponents' systems at baseband level by using a common transmitter, a common receiver, and inject controlled transmission impairments in the rf path." In a succeeding chapter there will be more on these tests as information becomes more specific and complete—but the form of summaries and explanations rather than a slavish quotation of every result.

Chapter 7

Cable Problems & Buzz

While not wholly report-related, these two generic system problems enter strongly into difficulties multichannel TV sound will have to face and overcome if the service is to become universally acceptable. Cable must overcome transmitter, oscillator, and separation snags, while receivers will have to respond to uncertainties in separate sound-video detection, stereo separation reduction, and increased noise floor measurements through cable systems. All this, of course, comes with the usual growing pains of a new technology that is only now reaching the market. Should any of these difficulties persist, the next generation of equipments will certainly reduce them substantially.

So bear with the industry as it delivers Zenith-dbx to a waiting nation with U.S. transmitters and domestic-designed and produced receivers plus a few Japanese decoder ICs. The next step is digital television, which is the subject of my forthcoming book in 1986. Then, even more topics of compression and expansion will be added to the electronic worlds of multiplexing, A/D and D/A converters, counters, and "menu" CRT displays. Obviously, consumer electronics is very much here to stay, even though there'll be a few hiccups along the way. The payoff, however, for both alert manufacturers and the consumers of video, audio, and satellite receiving products should be *tremendous*.

CATV PROBLEMS AND TESTS

There appear to be two prime problems in marrying cable television to TV multichannel sound: one is the proximity of immediately adjacent channels, and the other involves secure audio scrambling at multiples of the TV horizontal line scan rate. According to the National Cable Television Association (NCTA), multichannel signals (MCS) will cause the FCC-approved Zenith system decoders to switch into stereo even though only monophonic continues to be transmitted. In further system examinations, cable scrambling reduced stereo separation markedly, and even audio distortions rose, in some instances, to better than the "FCC's average THD of 3 percent for regular FM and television broadcasts." And these results developed from introducing MCS into scrambling systems only, and not headend processors, converters, and television receivers, which should cause further degradation in total harmonic distortion. A very good number for total THD would be *less* than one percent.

The reason for this separate CATV report resulted from early EIA tests which NCTA claimed "did not evaluate the impact of the wide deviation MCS signal transmitted by these systems on various parts of the cable system and the effect of the cable system on the MCS signal." Consultants Cablesystems Engineering Limited were then hired to do testing and "determine the interaction between cable and MCS systems."

The following, then, will constitute a report of a report, extracting pertinent information from the survey just as it appeared at date of issue, October 1983. All tests were conducted at Matsushita's Industrial complex, 9401 W. Grand Avenue, Franklin Park, Illinois and Rogers Cable-systems, Inc. Push-pull trunk and distribution amplifiers were used with no frequency offsets, and "one channel in the midband had the aural carrier modulated with a pseudo multichannel sound signal." The upper adjacent channel was then instrument-monitored for picture interference resulting from multichannel deviation and visually on several home subscriber television receivers. Tests began August 8 and ended a month later on September 15, 1983. Field tests lasted two days in July, 1983. The dbx, Inc. companding system was *not* examined.

Scrambling & Descrambling Effects

Four different scrambling systems were examined, all of which used sync pulse suppression, carrying coded reconstruct commands as amplitude modulation on the audio carrier.

System 1 was divided into two conditions. In the first the decoder was always operating and any AM appearing about the 15 kHz aural carrier would be "converted to amplitude modulation of the visual carrier." The second condition occurred when the descrambler only worked with a specific "tagging" trigger on the aural subcarrier. The visual carrier would not be affected unless this "tagging" signal was present.

System 2 featured pulse-type sync suppression, with one test at 6 dB and the second at 10 dB suppression.

System 3 also had pulse sync suppression but included "dynamic" timing for decoding.

System 4 obtained sync suppression with a sine wave, operating at double the 15,734 Hz line scan rate.

Aural carrier amplitude modulation appears as a prime bugaboo in today's scrambling systems when used in descrambling the decoder. Scramblers may also introduce some phase modulation of the aural carrier, and limiters in TV IFs can produce AM to PM conversion that will show up in detected audio and cause problems in multichannel sound.

To analyze the problem, a visual carrier was adjusted for a −8 dBm signal to the test demodulator; the aural carrier 13 dB below video; and then modulated with 1 kHz audio at a peak deviation of 25 kHz. Detected modulation reached a spectrum analyzer with a window of 1-100 kHz at 300 Hz resolution. This audio was peak referenced to the top horizontal line of the analyzer graticule its display recorded at approximately −18 dBV. All four types of scrambling-descrambling systems were then coupled between the signal generator and analyzer.

The block diagram in Fig. 7-1 appropriately illustrates the test setup showing video and audio test signals passing through their respective switches and MCS encoders, then into channel 3 exciters and scrambling networks. The combined output finds its way into the converter/descrambler being examined, a demodulater, MCS decoder, receiver(s) and the test equipment.

Analysis of baseband outputs showed Zenith with a stereo pilot signal at 15,734 Hz. Scramble systems have 15.734 kHz levels from 1 to 20 dB below stereo pilot, possibly false-triggering stereo decoders if such levels reach the pilot, as well as creating noisy mono and possibly poor stereo separation. Telesonics operates with an L-R subcarrier at 2½ times horizontal scan. Unfortunately, the scrambling systems generate components even two or three times line scan, therefore there must be a

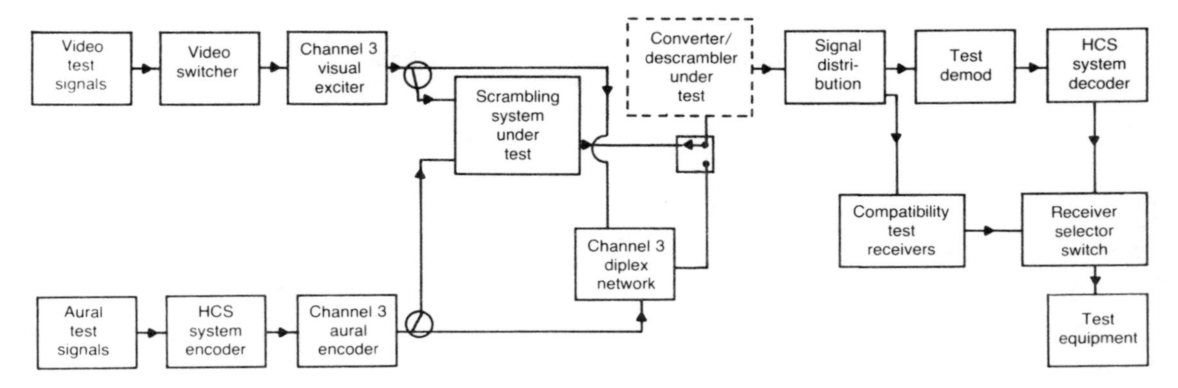

Fig. 7-1. Audio/video test signal setup for cable processing of multichannel sound (courtesy NCTVA).

notch filter in the stereo decoder to remove any 7.5 kHz whistle. In the EIA-Japan system, high frequency components shouldn't cause adverse effects since it is an FM-FM arrangement with the L-R subcarrier at 2 f_h.

Next, the input level of all descramblers was adjusted to +10 dBmV, and amplifier output after the descrambler adjusted to −8 dBm for the test demodulator input. With the aid of a specialized X-Y recorder, gain-phase meter, and Boonton signal generator, the four scrambling systems were checked for mono compatibility and "stereo upgradability." In other words, was mono adequate and could the examined systems accommodate good stereo? Demodulated outputs were channeled to each of the four systems under test and their stereo decoders to which a distortion analyzer and phase-gain meter were connected.

Multichannel sound was then introduced into each of the scramble/descramble systems, including separate SAP amplitude response and single tone distortion "with and without buzz pattern." Afterwards, test receiver baseband output was routed through a 15 kHz low pass filter and measured for single-tone distortion with and without SAP, followed by noise floor measurements for left channel, SAP and receiver compatibility. Radiometer BKF10 provided the swept test tone signal and H-P 3325A delivered the positive going ramp sweep from 20 Hz to 16 kHz in 99 seconds. Amplitude versus frequency plots were recorded for each left channel response.

Results showed no significant problems with left channel amplitude response among the scramblers. Stereo separation measured better than 30 dB for all systems except EIA-J *below* 60 Hz for system 1. But system 2 constricted EIA-J separation by between 20 and 26 dB; Telesonics from −21 to 25 dB; and Zenith by 30 to 40 dB down, and below the 30 dB level. But system 3 affected EIA-J by only 20 dB; Telesonics by 23 dB, and Zenith by only 12 dB, and system 4 showed even less separation problems than system 3, affecting Zenith by only −6 dB.

Single tone distortion versus frequency tests were instituted with the EIA buzz pattern and 5 degree ICPM and recorded for each of the left channel outputs. The test receiver first accepted quasi-parallel and then split sound modes.

Results are as follows: EIA-J distortion increased 0 to 15 dB; Telesonics increased 8-10 dB through its notch filter; and Zenith increased 0 to 10 dB—all in the quasi-parallel mode. Split sound detection increased distortion to the extent that its 5 to 20 dB would produce an unusable signal.

Noise floor measurements were next since they are unrelated to operating signals and indicate susceptibility to self-generated random noise, hum, and other undesirable disturbances. Once again, the test receiver operated in the quasi-parallel and then the split sound mode with EIA buzz going into the video exciter during tests.

Results showed that EIA-J rose 4 to 13 dB; Telesonics advanced from 0 to 5 dB in standard and

9-17 dB with notch filter; Zenith rose from 0 to 10 dB.

All systems showed a 20 to 30 dB increase during split sound applications.

Audio crosstalk from other system channels was the next objective (Fig. 7-2), with 400 Hz oscillators adjusted for 16.7 kHz deviation of the main and 10 kHz deviation of the SAP channel. White noise outputs were routed through EIA-J colored noise weighting filters, with noise generator levels adjusted 7.3 dB below reference tones. Black burst excited the video channel. Noise floors were measured initially with no modulation and, afterwards, with noise modulation selectively applied.

Results showed that mono crosstalk did not change appreciably when signals proceeded through the scrambling systems, and neither stereo nor SAP showed ill effects. In fact, the final rms dB figures are remarkably similar with Zenith possibly having the edge under some circumstances.

Carrier deviation in excess of 25 kHz could possibly produce sync "reconstitution" timing difficulties or apparent low frequency noise in the recovered signal.

According to NCTA's Multichannel Television Sound Engineering Report, high Q rf bandpass filters permit the aural carrier to be recovered with "minimum products from the adjacent video modulation." Aural FM falling on filter skirts will convert to AM in detection, so such filters must be able to pass multichannel sound with accurate AFT centering to keep picture noise at mimimum. Therefore, FM products must be suppressed 50-55 dB below threshold so that circuit aging and temperature changes will not develop this unwanted interference with extended equipment use. L-R above the bandpass filter can add to mono below the bandpass filter and increase noise modulation.

Equipment used consisted of an aural exciter, video reference input, the two types of scramblers and a descrambler, a TV demodulator and an oscilloscope, spectrum analyzer, and waveform monitor. Signals between 1 and 15 kHz were injected, with the visual carrier modulated with an IRE 100 level flat field signal at 87.5 percent modulation.

Results proved that "worst case" multichannel sound interference was greater than mono at lowest audio and more interference was visible in the pic-

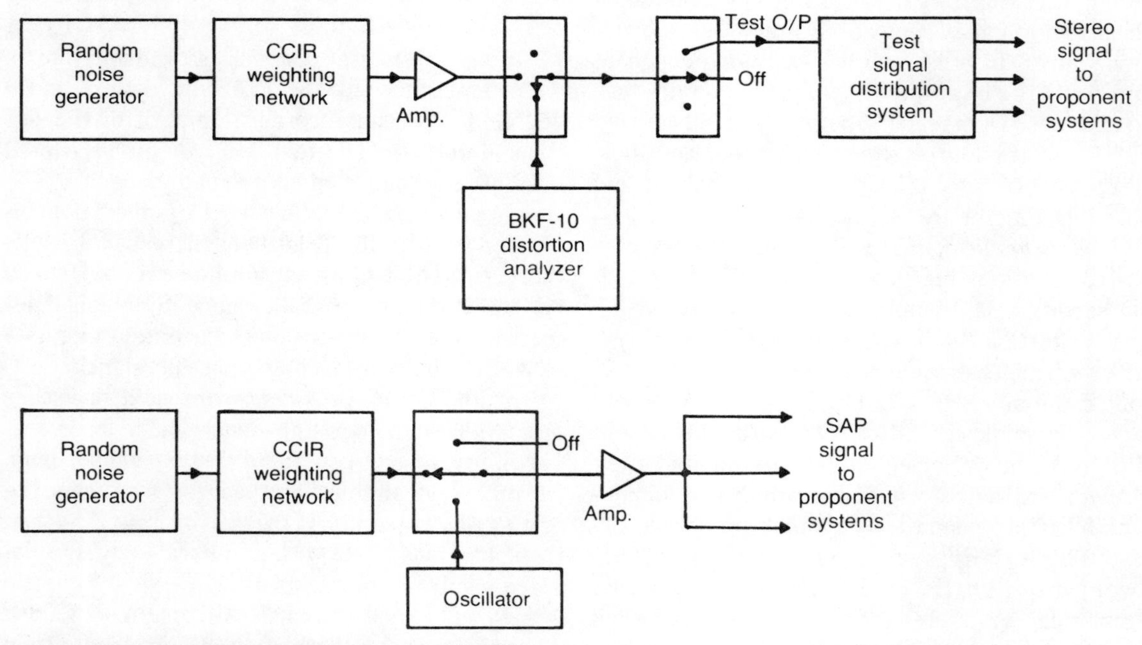

Fig. 7-2. Crosstalk evaluation setup (courtesy NCTVA).

ture at lower audio frequencies, even though there were differences among the three systems.

When subjective tests were recorded for multichannel sound effects on the visual display following descrambling, sinewave suppression systems produced picture degradation consisting of low frequency beats or noise varying with the audio. Pulse suppression systems were not affected, although there were constant small beats in the picture apparently originating from the test setup.

A group of television receivers with lower adjacent channel sound traps were also checked on the cable system to see if the MCS signal was sufficiently removed by the adjacent channel sound trap. Visual carriers were modulated with a 50 IRE flat field, SAP *on* but unmodulated, and channels 3 and/or 4 modulated with test signals swept from 20 Hz to 16 kHz. Receivers were "fine tuned" to minimize adjacent channel interference.

Results indicated less interference with MCS than with mono, but some receivers produced greater interference effects than others at higher video/audio ratios with MCS—a condition to be expected.

Head end processors consisting of strip amplifiers, demodulator/remodulators, heterodyne processors with notch aural carrier level control, and heterodyne processors "which separate the aural and visual carrier for processing." Aural bandwidths can reduce stereo quality, as expected. Signal levels were +10 dBmV for visual and 0 dBmV for audio carriers.

Results revealed the "stereo amplitude response was not significantly affected by the processors," even though separation did decrease by 5 dB max. with MCS signals, and noise floors rose to 25 dB after processing. SAP channel frequency response, however, did *not* change to any extent when processed, and noise floors rose from 3 to 16 dB. SAP crosstalk, however, did "increase significantly" in processor 3 and somewhat in processor 5 when excited by *all* MCS systems.

Cable channel converters usually used have varactor and not crystal-controlled oscillators. The former generates significant "phase" noise such that *separate* audio detection in TV receivers won't operate properly.

With visual carrier converter input set at +10 dBmV and aural carrier at 15 dB below visual, color bars were introduced into the visual exciter and left and SAP channels were measured.

Results produced additional stereo noise floors from 0 to 6 dB in quasi-parallel and over 40 dB in split sound detectors. SAP noise floors increased only 6 dB max. regardless of detection methods.

Baseband converters in demod/remod systems—up to peak deviations of 25 kHz—did not exhibit visible interference from adjacent channels with any of the proponent MCS systems.

Overall MTSER Conclusions

Multichannel sound signals delivered to cable subscribers would suffer degradation in separation, distortion, and noise levels, especially low frequency noise apparent in sinewave sync suppression systems. Some headend processors reduced qualities of MCS signals, but demod/remod equipment with 4.5 MHz aural carrier will usually pass MCS. Channel converters, however, do increase noise floors while baseband converters will pass only mono portions of MCS. Phase noise in varactor oscillators prevents use of receiver split sound detectors from operating satisfactorily on cable systems tested.

AUDIO BUZZ

Characterized as a nasty sound in any language, buzz (as in bee) pops up at the most inopportune moments, especially in poorly designed television receivers, to annoy and even ruin enjoyment of many music programs. In mono TV sound you've heard it all too plainly when poorly-controlled TV stations permitted video to ride above +100 IRE, especially on commercials, with resulting raspy, low frequency, scratchy buzz.

Technically, all this occurs when phase-modulated portions of the video carrier enter the sound channel, producing the disturbance which emerges during the normal blanking interval as vertical sync. In mono sound and at limited low and high frequencies, buzz lasts only a short time and may not be that unpleasant in the better-designed receivers. But when stereo or multichannel sound combines with subcarriers, the problem is compounded and becomes quite prominent.

Carl Eilers and Pieter Fockens of Zenith Electronics Corp. recognized the problem even before 1981 and produced a paper on the subject for the IEEE Transactions on Consumer Electronics in August of that year. They defined buzz "as the result of video-related incidental phase modulation (IPM) of the intercarrier," and concluded that the sources were nonlinear parametric and transit time effects, and linear AM-to-PM conversions. The former develops in broadcast transmitters and the latter in receivers. Like the 6 dB/octave increase in FM thermal noise, this same factor "is also active" on buzz since peak angle deviation at double the frequency produces a proportional FM detector output. This, of course, is all 4.5 MHz intercarrier FM modulated sound, which can also pick up buzz from reverse mixer feedthrough, multi-tone intermodulation and CATV (Cable TV) processing. The sources, therefore, are not simply overmodulation, but originate from a variety of highly technical causes all potentially damaging to multichannel sound.

Intercarrier sound, itself, originates from the 4.5 MHz difference between the television intermediate frequency (IF) video carrier at 45.75 MHz and the 41.25 MHz sound carrier. Developed principally by General Electric, tuned circuits accept this frequency difference, amplify, detect and deliver the FM information to demodulators, final amplifiers, and speaker loads. A separate detector in color receivers is required to overcome problems with the color subcarrier at 3.58 MHz producing a 920 kHz beat with the FM-developed 4.5 MHz carrier. Initial detection occurs prior to the receiver's video detector since, at that point, the original 41.25 MHz sound carrier has been eliminated by tuned traps. In integrated circuits, quadrature-type demodulators are now quite popular and deliver an audio output voltage proportional to input frequencies. Unfortunately, sound channel limiting doesn't always remove vertical blanking information and, therefore, you sometimes hear cross modulated sound with sync and video.

The simplest sound detector is a diode, which is a half wave slope or envelope detector. The video carrier, being amplitude modulated, cannot pass the FM limiter. But phase angle modulations may filter through and if they "are in the audio frequency range or in the composite baseband range of frequencies for multichannel sound they become audible," according to Fockens and Eilers, and the sync and blanking portions then produce buzz.

Synchronous wideband detection also passes incidental phase modulation (IPM) by another means, even though the results are equivalent. With relatively narrowband audio, the wide bandwidth reference carrier may be twice as wide, and IPM slips through even in multichannel situations. If the reference information (channel) is a few kilohertz wide, IPM is blocked above those frequencies and won't pass into the sound channel.

Other factors noted include study of equal incidental phase modulation in both video and audio carriers. A diode detector will eliminate most IPM from the intercarrier because of a "differencing" of the two carriers, and the same applies to a synchronous detector. In the days of RCA's and Dumont's early receivers, sound was taken off shortly after the tuner, a discriminator did the demodulating, and video and audio were, consequently, processed independently. True, additional stages of amplification were needed, station selection tuning and alignments were considerably more difficult, but there was no sync or overmodulation buzz because there was *no* incidental phase modulation. So, as we are to see, video-audio separated sound processing has, once again, become attractive "provided no tuner IPM is introduced," presumably from automatic frequency controls (AFC).

What effect frequency synthesis and phase-locked loops may ultimately produce will be discussed when the new receivers are reviewed and their stereo circuits analyzed. But we do know that synchronous detectors, like envelope detectors, will lose their reference in instances of overmodulation, but in a different way among intercarrier conventional receivers. In envelope detectors, when the visual carrier, according to Mssrs. Fockens and Eilers, closes toward zero, the 4.5 MHz sound detector beat disappears and noise rises audibly as a "rattle" or buzz. Conversely, with visual overmodulation, the synchronous detector loses its reference because of interruption of the 45.75 MHz clock reference carrier with the same results. Fortunately, this does *not* hold true for split soundvideo receivers since incidental phase modualtion has already been eliminated.

Among transmitters—aside from obvious overmodulation—there's no (or shouldn't be) IPM in the audio carrier since audio and video are not combined at rf until the passive diplexer and actual antenna. However, a situation may arise, according to Fockens/Eilers, in which small signal phase varies with bias and undergoes differential phase distortion "while the luminance-modulated carrier experiences IPM." Such nonlinearity, they say, results in harmonic, intermodulation, and differential gain distortion, "but not per se, differential phase distortion." Parametric or transmit time effects are prime causes involving junction capacitance, dynamic resistance and transit times in Klystron amplifiers.

Incidental phase modulation (IPM), nonetheless, does vary in direct proportion to carrier level, and signal-to-buzz ratios often are double that of modulating frequencies, rising in one Nyquist slope analysis 12 dB/octave versus 6 dB/octave for IPM. Authors of the paper say that transmitter IPM should, therefore, "be limited to 1 degree or preferably less."

IPM, by the way, may be measured with a television demodulator in synchronous detection with both in-phase and quadrature detectors. Inphase modulation is ignored as phase modulation is checked on a staircase video signal (Tektronix 1450 demodulator). Other important parameters are: audio frequency response, SAP crosstalk interference, noise floors, channel separation, peak deviation of aural carrier, main and stereo subcarrier, visual carrier, and signal-to-buzz ratios for mono, stereo, and SAP modulation. Adjacent channel crossmodulation is another problem to watch for; as is harmonic distortion resulting from overloads.

There is also a proposed rule mandating 3 degrees maximum IPM of the visual carrier between reference white and blanking levels, and 5 degrees max IPM from blanking level to sync tip. These probably will also require transmitter modifications or adjustments, depending on the equipment at hand.

CATV multichannel sound problems—already considered in detail—are also said to be a prime difficulty; so much so that the Federal Communications Commission has deferred any ruling on "mustcarry" cable television because of carrier separation problems. Video/audio broadcast ratios are normally 7-10 dB, while CATV operates between 15 and 20 dB. Therefore, CATV has an 8-10 dB greater sensitivity or possibility of buzz, making it very difficult on multichannel receivers. CATV up/down heterodyning could also become a problem among these sets.

We'll try to have some of the transmitter "cures" spelled out in Chapter 9, and receivers should really tell their own stories via circuitry and a few good, explicit technical papers. In addition, you'll receive some intensive material on noise reduction systems which are a very important part of this entire multichannel scheme. And as Mssrs. Eliers and Fockens strongly suggest, "buzz is a systems problem. It will take cooperation between transmitter manufacturers and operators, cable equipment manufacturers and cable system operators and receiver manufacturers to bring it down to acceptable levels—if not eliminate it."

You now know why we have devoted this much space to a single subject—even though the patient isn't terminally ill, a complete cure won't be easy despite today's excellent consumer design engineers and advanced integrated circuits. So don't expect perfect systems to emerge across the market immediately or inexpensively. There'll be a great deal of engineering time to devote to the problems before general satisfaction is achieved. In the meantime, be prepared to pay a premium for initial receivers, then much less once manufacturing settles down and licks the various glitches that crop up here and there. Enjoy what wasn't there a few months ago—all in the name of progress!

Chapter 8

The FCC's Final TV Actions and Approved Regulations

After all the years of proposals, tests, investigations, and responses, the Federal Communications Commission on the afternoon of March 29, 1984 finally gave its unanimous approval to: "The use of subcarrier frequencies in the aural baseband of television transmitters." (Docket No. 21323, RM-2836). Unlike AM stereo, one system was permitted under certain conditions even though others were *not* excluded. Translated, this means that the Zenith-dbx system is mandatory if the stereo carrier appears 4.5 MHz from the video carrier at a fixed frequency of 15,734 Hz—the same as all color horizontal scan frequencies.

If others wish to broadcast, then they will have to choose some competing subcarrier that is different from this particilar frequency but within FCC guidelines. Since the Electronic Industries Association chose the system following exhaustive testing and recommended *only* Zenith-dbx to the FCC, there'll probably be a long wait for any serious competition to appear, in at all. This also means you'll have stereo TV sound on the air before 1985 and receivers with at least two multi-channel versions by the fall of 1984. So instead of a four-year wait for industry's "open market" shakedown—as in the case of AM stereo—the competition is burning midnight oil getting products ready for coming winter sales and subsequent consumer enjoyment.

The FCC did not take its step-by-step decisions lightly; rather, the pace was very deliberate in both its Second Report and Order of April 23, 1984 as well as the Further Notice of Proposed Rule Making adopted July 28, 1983, and released August 15.

In the Further Notice a number of inquiries were received from broadcast licensees about using TV aural baseband subcarriers for programming to special interest groups. The FCC noted that present rules prohibit even the SCA (subsidiary communications authorizations) already enjoyed by FM. Consequently, the Commission considered that TV aural subcarrier operations "may be unnecessarily restrictive." The Commission noted that on June 30, 1981 it established new rules permitting TV subcarriers for ENG cueing and coordinating and acceded to suggestions that subcarrier injection levels be determined by whatever means the licensee desired. But the injection level of any single subcarrier must not exceed 15 percent nor could the injection level of all exceed 15 percent. Further, system AM and FM noise levels were limited to −50 and −55 dB, respectively, with crosstalk also limited to −55 dB. However, at that time, it did reject the proposal to expand maximum deviation above ±25 kHz since that parameter was still under study.

Members also knew that current technology now allows aural baseband into many non-broadcast related uses such as paging, electronic mail delivery, fax., and traffic light and sign control. There are 1,181 television stations now broadcasting, of which 896 are commercial and 285 non-commercial (in 1984). "Each of these stations could be providing two or more subcarrier services . . . (and) thousands of hours of subcarrier services could be offered at virtually no technical cost." But they did think that in the beginning, television with stereo programming might be limited—a point well taken even three years in advance of final approval since transmitter hardware is just being made available in quantity during 1985.

Further, it's safe to say there will be many more stereo-receptive receivers available than proportional numbers of transmitters to excite them. We'll have considerably more to say about this aspect in the transmitter-exciter chapter where the newest designs and their adaptations to old and new transmitters will be undertaken. Unfortunately, just plugging in an extra IC or two is only part of the answer. Then there'll be some transmitters that just won't adapt at all, and some station owners who won't provide the money needed to make the change. So FCC foresight merits a 10 in this instance even though some of its other prognostications and decisions may not be imbued with Solomon's wisdom or David Sarnoff's clairvoyance.

Citing the Public Broadcasting Amendments Act of 1981, the FCC also thought that public broadcasters should be allowed "a full range of services on their aural basebands"—a consideration which may be unusually beneficial soon since the 80-station public broadcast network will probably be the first to offer stereo on-the-air programs to at least some segments of the population because a few have already been working with the medium over the past year or two. Public broadcasters are also receiving less and less money in Federal funding, and therefore must "generate substantial sums of additional revenues from the pursuit of commercial activities," according to the U.S. House Committee on Energy and Commerce. Section 399B of the Communications Act states that "each public broadcast station shall be authorized to engage in offering service, facilities, or products in exchange for remuneration."

The Commission believes that TV aural subcarriers should be considered ancillary broadcast services and regulated accordingly under FCC regulations, part 73. Should the broadcaster wish to offer either common or private carrier services over its TV subcarriers, the appropriate common or private carrier regulations should apply. However, the treatment of subcarrier operators would vary somewhat since they would *not* require technical approval of a total TV broadcast station. But TV aural subcarrier use would be considered by the FCC "as a secondary privilege that runs with the primary television broadcast station license, and a broadcaster using a subchannel for private or common carriage would remain a broadcaster for all other purposes." Only subchannel use for nonbroadcast-related purposes would be regulated as private radio or common carrier.

Among technical considerations, Commissioners then recalled their "free market" approach to subcarrier use and said that while "minimal" performance standards have traditionally applied to telecommunications under consideration, they considered that such "regulation (should) no longer be necessary," and suggested that aural subcarriers be governed only by whatever technical rules are needed "to ensure the integrity of primary visual and aural service, and to preclude interference to other licensees." As an example, the Commission cited 20 dB as sufficient left and right channel stereo separation, but 10 dB was probably not. Therefore, they argued, minimum performance standards are "one means of assuring this result."

PROPOSED TECHNICAL STANDARDS

Still in the July 1983 time frame, Commissioners decided this was a good opportunity to put forward some definite parameters on multi-channel sound since final approval of the system could not be too many months away—conjecture that became reality within eight months, virtually to the day.

They wanted all subcarrier transmissions to deliver a composite stereo signal "fully compatible" with existing mono receivers, and that main program sound levels should not be significantly affected. Consequently, they wanted no more than 2 dB deviation of the aural carrier by the main channel signal and to continue 25 kHz deviation of the main channel but increase deviation of the main carrier to "accommodate" extra multiplex subcarriers. The Commission was unsure of what this would do to television receivers, so members requested substantive comments from industry "to assist us in establishing a new maximum permissible deviation."

Next the problem of visual and main channel aural interference was considered, and it was decided that crosstalk between channels was the most likely source. Therefore, they proposed a 60 dB attenuation of any non-stereo subcarrier signal into the main channel, and any crosstalk from the stereo subcarrier into the main channel should be restricted to −40 dB. At the same time, they recognized that industry was thinking of a stereo subchannel between 16 and 55 kHz, but they saw "no reason to restrict it to that range" and proposed to permit stereo subchannel operation "anywhere within the usable aural baseband," although an upper limit of 120 kHz was thought to be reasonable. After this, they recommended a 40 dB attenuation of any signal outside the aural baseband, and then suggested a pilot tone for receiver recognition and control with no limit on "the number of pilot tones or their uses," although they should be restricted to 15-120 kHz in the aural baseband. It was also recognized that power levels for aural carriers versus video carriers might be increased to their former levels of 30-50 percent for expansion of subcarrier service area coverage.

The Commission also recognized that cable television systems and subscription television systems might be affected by any multi-channel TV sound authorization. Indeed, there was a letter from Blonder-Tongue Laboratories saying that subcarrier protection of STV audio "could be defeated by some of the proposed multi-channel sound systems." The FCC replied that its initial reaction wasn't intended to "protect the operations of a few subscription TV stations at the expense of the general public benefit, should it turn out that multichannel TV sound systems incidentally detect certain subscription audio signals." What's more, the FCC remarked, interception of subscription TV audio by multi-channel sound decoders wouldn't become "a fatal breach" of subscription TV security since video would remain scrambled.

Adjacent Channel Cable TV Problems. FCC engineers did take a considerably more serious view of adjacent channel cable TV problems if total aural baseband subcarrier deviation goes up 75 kHz. They wanted to know what kind of modifications would be needed in cable for quality multisound reception, as well as the cost. They wished comments on whether policy should require CATV to carry multi-channel sound services. Among others, the cable industry went into extensive testing and you'll see the results later in a separate discussion of this problem—which is so severe, the FCC has deliberately delayed a final ruling on its "must carry" regulation so the matter can be studied at length before final determination. Resolution of this issue may take a number of months or even a year before it's settled. You'll see why in the CATV sound explanation.

The FCC's objective in this Further Notice of Proposed Rule Making, dated July 28, 1983, was to "fully expand the services permissible on TV subcarriers by removing its present limitations." They foresaw many cost-competitive alternatives for many services now excluded from using the TV aural baseband. A substantial number of small businesses, they thought, might be positively affected, but some negatively. The former include small commercial TV stations and businesses furnishing "previously precluded competitive services and equipment suppliers." Those adversely affected might include commercial and non-profit businesses now current users, FM subcarriers, and service suppliers of other transmission methods which would lose income to newly-approved competitors.

That's the extent of what the FCC said on July 28. From here on we'll go to the final ruling and see how all the foregoing turned out. There may or may not be surprises for some of you and considerable for others, depending on how you straddle the regulatory fence. At least you're aware of the procedure, its extent, and nation-wide participation by many segments of the industry.

THE FCC'S FINAL RULING

With all Commissioners present and voting, the four members and Chairman Mark Fowler voted unanimously at 4:15 P.M. March 29, 1984 to authorize use of subcarrier frequencies in aural basebands of TV transmitters. Commissioner Rivera dissented in part, and Commissioner Patrick concurred in part. But the Zenith-dbx system was approved, and a new industry born throughout the United States. At last, America was to have stereo TV sound and all that went with it. What follows now is the decision's background and some adopted regulations to go with it.

Reasons and Special Rulings

The particular proceeding originating in 1977 when Boston Broadcasters, Inc. asked to use a TV aural baseband subcarrier for cueing and coordinating electronic news field personnel. This resulted in a Notice of Inquiry dated July 1, 1977 as to the possibility of TV stereo sound, bilingual programming, and augmented audio for the blind. Enthusiastic comments were received, especially over the use of TV baseband for operational convenience. The Commission, accordingly, adopted rules on June 30, 1981 permitting limited use of this aural baseband for electronic news gathering (ENG) but, at the same time, delayed action on a request for maximum aural carrier deviation to be increased above the ±25 kHz limit.

While this was going on, the Electronic Industries Association (EIA) and its Broadcast Television Systems Committee (BTSC), organized a Multichannel Sound Subcommittee (MSS) to study the possibilities of enhanced TV sound and suggest relevant technical standards to the FCC. Involved were: a stereophonic sound channel, a second audio program channel (SAP), and additional multipurpose subcarriers. In December 1983, according to the Commission, the Subcommittee selected one multichannel sound transmission system and one audio companding noise reduction system from among the six competitors. Winners were Zenith and dbx, Inc. Combined, these are now known as the BTSC system for the committee that chose them.

The FCC also recalled its Further Notice of Proposed Rule Making of July 28, 1983, (already discussed in detail at the beginning of the chapter), and undertook to supply many considerations and answers. Many respondents, the Commissioners reported, were critical of the FCC's "marketplace" approach in AM stereo and that "there would be *no* benefit in adopting general technical standards for MTS . . . (since) such action would likely result in marketplace uncertainties and inaction." But they did want specific system standards for TV receiver compatibility to protect the public's investment.

Cable operators conceded there should be specific system standards but were unhappy with all EIA-tested packages because "of increased aural bandwidth requirements." Commentators included Blonder-Tongue, Grumman Aerospace, Time Period Modulation, Duncan Laboratories, Rocktron Corp., Alpha-Omega, and Telesonics, one of Zenith's competitors, who wanted a further continuation of the proceedings.

The Commission was happy with EIA's specific MTS proposal, but conceded that technology will continue to advance rapidly and did *not* believe it should be restrained. It also liked the idea of a pilot subcarrier at 15,734 Hz—television's horizontal scan frequency—but moved, of course, 4.5 MHz away from the video carrier where there would probably be little interference if audio and video are properly separated. This, of course, triggers receiver stereo circuits and alerts the operator to TV stereo reception.

Commissioners reasoned that if the BTSC system uses the 15,734 Hz pilot subcarrier it would protect BTSC receivers from operating on other formats, and allow other multichannel sound transmission systems to operate, based on marketplace demands. They also decided that no other MTS system should offer a pilot subcarrier between 16 kHz and 120 kHz, ±20 Hz to modulate the aural transmitter more than ±0.125 kHz. They considered that the 40 Hz window would offer sufficient protection for "state-of-the-art tone detection circuitry in BTSC receivers." Above 120 kHz the FCC specifies 40 dB of attenuation to prevent any interference with video.

MTS channel crosstalk (except the stereo difference channel) was scheduled to be limited to −60 dB, and that of the proposed difference channel crosstalk into the main channel by −40 dB. Although EIA sanctioned these values, the FCC said this should be left up to the marketplace to determine the "balance between listeners needs and crosstalk limits." And so members declined to adopt any specific standards. But it was recognized that MTS would result in deviation over and above standard ±25 kHz, and so "an additional 50 kHz deviation" was allowed for MTS, "to be allocated according to the needs of the specific system." This, then, permits a total of ±75 kHz deviation for MTS. And any MTS that offers piggy-back stereo difference information on the visual carrier will be permitted 50 kHz "additional" deviation, for a total of 75 kHz. As to left and right channel separation, the Commission expects that stereo systems meet L−R audio performance standards, but won't require a 30 dB regulation since it is considered too "restrictive."

Commissioners also believed public broadcasters should be allowed to render subcarrier "services" either commercially or noncommercially to aid in increasing revenues for their support. At the same time, broadcast licensees are permitted to decide "which TV audio subchannel service to offer." Further, separate subcarrier transmissions not connected with program content "do not warant the protective regulation accorded... primary broadcast services." In the matters of FM and TV subcarrier common carrier or private carrier regulations (if invoked), and "restrictive or exclusionary" entry regulations, the Commission said this would be decided for both services in BC Docket No. 82-536. As for the Fairness Doctrine and political broadcasting requirements, the FCC believes that the "statutory requirements of reasonable access and equal opportunity are adequately satisfied by permitting federal candidates access and opportunity on the licensee's regular broadcast operation and does not require access to ancillary services." So aural TV subcarriers are *not* subject to either the Fairness Doctrine or Sections 312(a)(7) and 315 of the Communications Act.

Cable Mustcarry

This has been a sticky problem for the Federal Communications for both technical reasons and contesting viewpoints from other services. Broadcasters would have stereo sound regardless and say "it is technically feasible for cable systems to do so." Other comments say a mandatory requirement should apply to multichannel sound as it does to the chroma-suppressed 3.579545 MHz subcarrier. MST, NBC, and PBS all believe cable should have "all aural subcarrier signals" so that subcarrier services could be developed rather than discouraging manufacturers in producing quantities of this equipment. Further, MST states that EIA tests show "the vast majority of cable systems can carry multichannel sound services *without* degrading present service or causing inter-channel interference." Any cable systems that can't retransmit aural subcarriers on the main TV channel should be required to offer multichannel sound services over vacant FM radio channels "or through other means if they provide such services for their nonbroadcast or pay services."

The National Association of Broadcasters (NAB) asserts that multichannel sound over most cable systems is automatic, does not burden CATV systems, and needs no extra "steps" by cable operators. The Association does note that some headend equipment may require adjustments, modifications, redesign, or replacement for acceptable stereo performance, but any adjustments are "relatively minor" and do not involve substantial costs.

But it does recognize that many settop converters now used by some CATV systems are actually incompatible with multichannel sound, but says such converters were manufactured when the principles of multichannel sound were well known. NAB and other objectors, want anti-stripping regulations imposed immediately on cable "capable of passing multichannel sound signals and on systems otherwise providing stereo or SAP for *any* programming."

Cable systems, for their part, say they're not ready to carry this new development, and the National Cable Television Association says the industry "will likely provide multichannel sound where it is technologically feasible "but what the decision . . . should be left up to the cable operators." Heritage Communications suggests that if there is a demand for multichannel services, cable TV will provide them, notwithstanding the absence of requirements to do so. But if no demand exists, Heritage thinks cable should not be obligated. NCTA says that after extensive tests, some CATV systems could offer multichannel sound without serious degradation but other systems "will encounter substantial and unacceptable interference." Other CATV comments say that policies for standard "mustcarry" rules don't apply to multichannel sound since broadcast stations were protected from cable initially, and the same conditions now do not exist. Later, the FCC will issue a "neutral" Notice of Proposed Rulemaking "to further explore this matter."

FCC REGULATIONS

Published in April 1984, OST Bulletin No. 60, entitled "Multichannel Television Sound Transmission and Audio Processing Requirements for the BTSC System," and are the FCC guidelines "for stations employing the BTSC system of multichannel sound transmission and audio processing." In short, these are the general, technical rules that apply to TV stereo, second language programming "and any other broadcast or non-broadcast use." Stations with emissions at 15,734 Hz are required to comply. A list of appropriate definitions and rules, as issued by the FCC, is included at the end of the book in Appendix D.

Chapter 9

Multichannel Sound Transmitters

Since the BTSC (Broadcast Television System Committee) system consists of the Zenith multichannel TV sound system and dbx companding electronics, it's only fitting that we begin these two final chapters with introductions from Zenith, and then progress to specific equipments that will actually be operating in both broadcasting stations and consumers homes. In this way we can probably offer a better general understanding of the ultimate product. Precise information may be somewhat scanty due to on-going development and the usual production changes involved with any new undertaking, usually regardless of source or function. In any endeavor there are always compromises, with some being more advantageous, acceptable, or cost-effective than others. At any rate, all follow the Zenith-dbx lead as specified by BTSC and the Federal Communications Commission, since at this point there is no competition. So let's take both Zenith and dbx system projections, initially, and work as best we can from there, using these as fundamental generics, with RC and IC electronics to come.

We might add there is already additional work under way toward quieting the old nemesis "buzz" since dbx only operates on L−R (the stereo portion) and not on L+R (the mono portion). So with larger speakers, infinitely better audio bandpass and improved receiver amplifiers, spurious audible instrusions of almost any description are magnified and heard a significant distortions. Both EIA and FCC, of course, continue to be immediately involved, as do fellow engineers in this country and abroad. You might also like to know that Panasonic (Matsushita) and NEC remain the prime sources for dbx's companding ICs, at least in 1984. With sufficient demand, of course, that could change and second sources would, undoubtedly, be licensed promptly.

ZENITH TRANSMITTER SPECIFICATIONS

Probably the best way to go about this is to extract several illustrative figures from Carl Eilers' (Zenith) latest (1984) report and proceed in this way, trying not to sound more than necessarily redundant in further describing prior material. Some aspects, nonetheless, will have to be repetitious because of the single, specific system involved—although how the several manufacturers proceed in putting hardware together could become very interesting indeed. We do understand, however, that adequate broadcast transmitters won't have a great deal of difficulty in converting to stereo sound once they receive the necessary equipment. There will be others, naturally, that can't convert at all—at least successfully—so expect somewhat slow, but steady progress, rather than a tornadic rush. The effects of consumer costs and broadcast equipment availability are going to restrict immediate, unlimited growth, especially in the "boonies." Cities like New York, Chicago, and Los Angeles should be on the air with good programming even before publication of this book. Less competitive areas won't move as fast, probably for several years.

In Review

Recall that in multichannel TV sound, main channel modulation is L+R with pre-emphasis of 75 μsec and peak deviation of 25 kilohertz. When L−R is encoded with dbx and L and R are statistically independent, main and stereo subchannels are peak deviated to 50 kHz. When level encoding is temporarily replaced by 75 μsec pre-emphasis, subchannel peak deviation is also 50 kHz. And when L and R are not statistically independent or L+R and L−R do not produce the same pre-emphasis, main and stereo subchannels are restricted to 50 kHz. See Fig. 9-1.

The second audio program (SAP) is also dbx compressed, operates at $5f_h$ (f_h = 15,734 Hz scan frequency), peak deviates at 10 kHz but deviates the main carrier by no more than 15 kHz. Then, of course, there's the pilot signal at f_h that may not deviate the main aural carrier by more than 5 kHz. Finally, comes the professional channel at 6.5 f_h, or 102.271 kHz that is specified for subcarrier and peak carrier deviations of 3 kHz. Second program and professional channels are straight FM, while the stereo subchannel is AM and double sideband with suppressed carrier. That pretty well covers what we would call the Stereo and SAP Service, Program A.

In Program B, there is stereo service, but no SAP channel and a considerable space (47-120 kHz) for professional services at the various 4, 5, 6, and 6.5 intervals indicated. As you can see, there's been no change in purely stereophonic service.

In Program C, however, the L+R main channel becomes available for monophonic service, with 2 and $3f_h$ available for professional channels, in addition to SAP and another professional channel at $5f_h$ and $6f_h$, respectively.

Where there's monophonic service only—as in Program D—you find a great deal of available space for professional channels, but no L−R stereo, or SAP included. The term "Program" is simply our designation for the entire frequency and bandpass allocation for multichannel stereo and services as identification for the several phases of this service as outlined so competently by Mr. Eilers. Note that pilot signals are prevalent in the two initial stereo modes only, not at monophonic, and that main channel frequency range extends from 50 Hz to 15 kHz.

Therefore, if you add 5 kHz for pilot, 50 kHz for stereo, 15 kHz for second program (SAP) and 3 kHz for professional, the total aural carrier peak deviation amounts to 73 kHz. As you see, this does *not* include the usual 25 kHz for monophonic since the main channel is limited to a total of 50 kHz under any conditions. A comprehensive parameter listing prepared by Carl Eilers appears as Table 9-1. To some these will appear as little more than dry statistics, but to engineering they should be pretty much law and ought to be adhered to for uniform broadcast standards. Naturally, there may be changes later on, but for now, what you see is what Zenith says you get. The implication could hardly be plainer, including specifications for dbx compression which, by the way, is used in both stereo *and* second program (SAP). Where dbx is *not* used, mono has 75 μsec pre-emphasis and voice on professional channels from 0.3 to 3.4 kHz has 150 μsec pre-emphasis. In the 0 to 1.5 kHz modulation frequency range there is neither pre-emphasis nor dbx since this is usually data transmission by frequency shift keying (FSK).

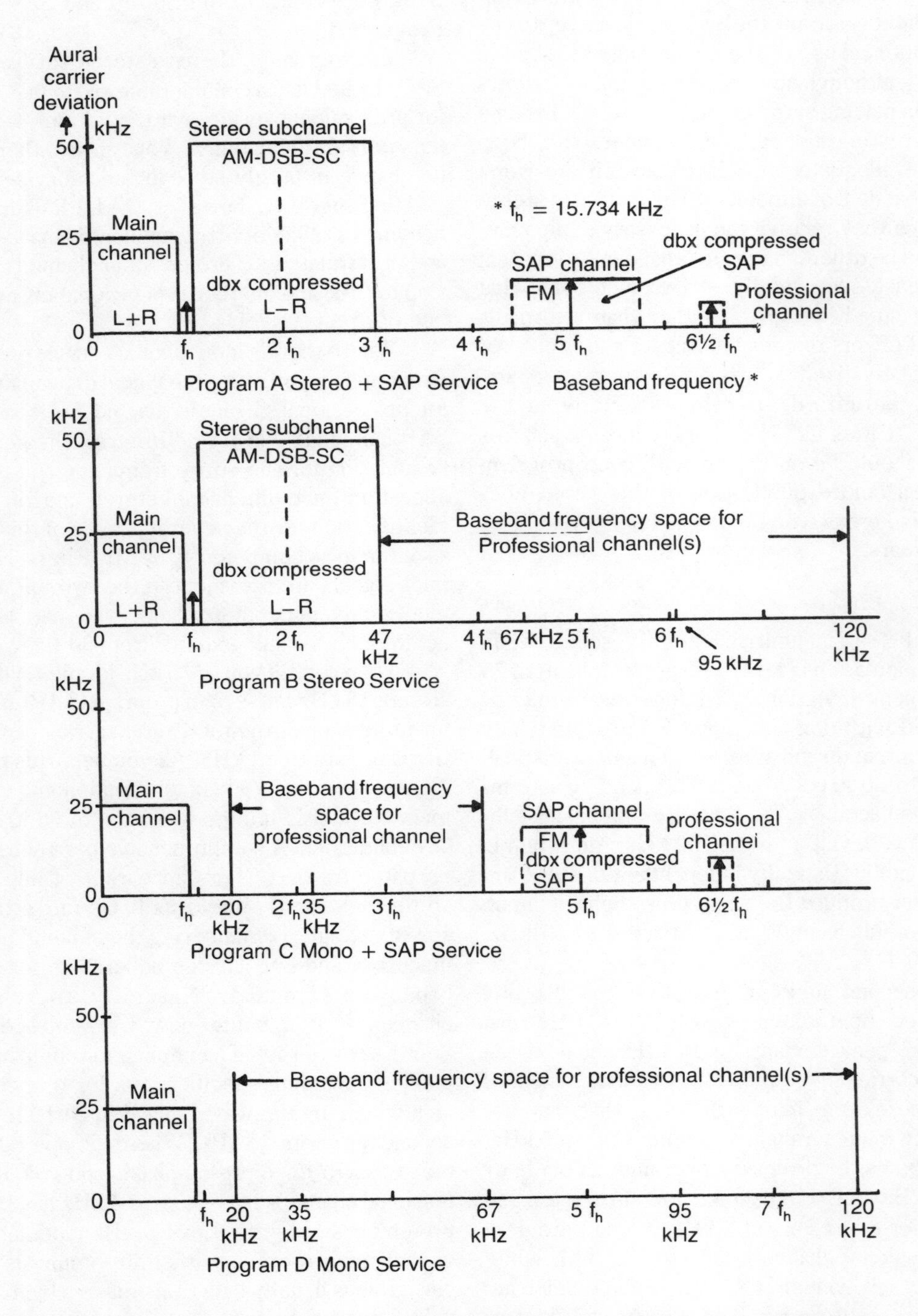

Fig. 9-1. Programs A-D illustrate various types of multichannel services (courtesy Zenith Electronics).

Table 9-1. Radiated Signal Parameters and Tolerances.

	Main Channel		
1.	Modulating Signal		L+R
2.	Frequency Range		50 Hz – 15 kHz
3.	Pre-emphasis		75 sec
4.	Aural Carrier Deviation by Main Channel	max.	25 kHz
	Pilot Subcarrier		
1.	Frequency (color program)		f_h = 15.734 kHz
2.	Frequency (black-white program) (no burst)		15,734 ± 2 Hz
3.	Aural Carrier Deviation by pilot subcarrier		5 kHz
4.	Pilot-to-Interference Ratio (1,000 Hz band) (Reference 5 kHz deviation)	min.	40 dB

Multisound Generation

With Mr. Eilers' expert help, let's refer to Fig. 9-2 and work through the system with somewhat different detail than has been recorded thus far. In addition, there'll be a block diagram of the dbx encoder so you can see how it fits the general and specific picture. This is very important since stereo separation is said to be directly influenced by dbx efficiency and compatibility in compression and expansion, meaning that transmitters and receivers should fully complement one another in all possible respects. Stereo noise, of course, is the main reason for dbx in the first place. But it does have separate stereo difference and SAP compressors in the transmitter and only one unitized expander in the receiver; consequently, maximum efficiency becomes quite significant for both radiator and receptor. That may also indicate some television receivers may reproduce stereo and SAP considerably better than others. Transmitters carefully retrofitted and with sufficient aural passbands should deliver fairly uniform signals.

Stereo Generator. The generator illustrated (Fig. 9-2) is said to be an example of those used in the FM stereo broadcast industry. Such generators, according to Mr. Eilers, are ordinarily of two types: one operates on time division multiplex (TDM) and the other on frequency division multiplex (FDM). In TDM, stereo right and left inputs are commutated (timed interruption of forward current by switching) through the generator at some subcarrier rate. This indicates bandlimiting and must be compensated for by bypass additions, increasing subcarrier levels, and/or cross-coupling control of inputs before commutating.

In frequency division multiplexing (FDM) each information channel occupies a specific portion of the transmitted energy and a number of channels may be carried over an appropriate system. Single sideband, suppressed carrier is a good example that comes to mind in two-way radio. Here, in a stereo encoder, matrixing and subcarrier generation are totally separate and level changes for main signals or subcarriers are only minor problems. The MTS stereo oscillator, however, locks to horizontal pulses taken from composite video. Such an oscillator, according to Mr. Eilers, could be converted to a very stable phase-locked loop with vco.

Normally, FM radio broadcast generators have lowpass filters and pre-emphasis circuits at their inputs. Note that they have been moved to positions following the equalizer and dbx encoder, so their outputs flow directly into eleventh order Cauer lowpass filters supplying the L and R matrix. The function of these filters is to path-protect the pilot stereo indicator and prevent crosstalk between mono L+R and L−R stereo channels. Although the usual FM stereo generator's composite lowpass filter has a passband of 53 kHz and TV stereo requires 46.5 kHz, stereo subchannel injection, SAP overlaps more third harmonic than does SCA, and needed SAP protection increases the stopband loss of this filter. Further, FM loss poles are at 19 kHz and should be at 15.734 kHz for TV stereo. Modifications are also required for conventional SCA generators, even though they are similar to SAPs.

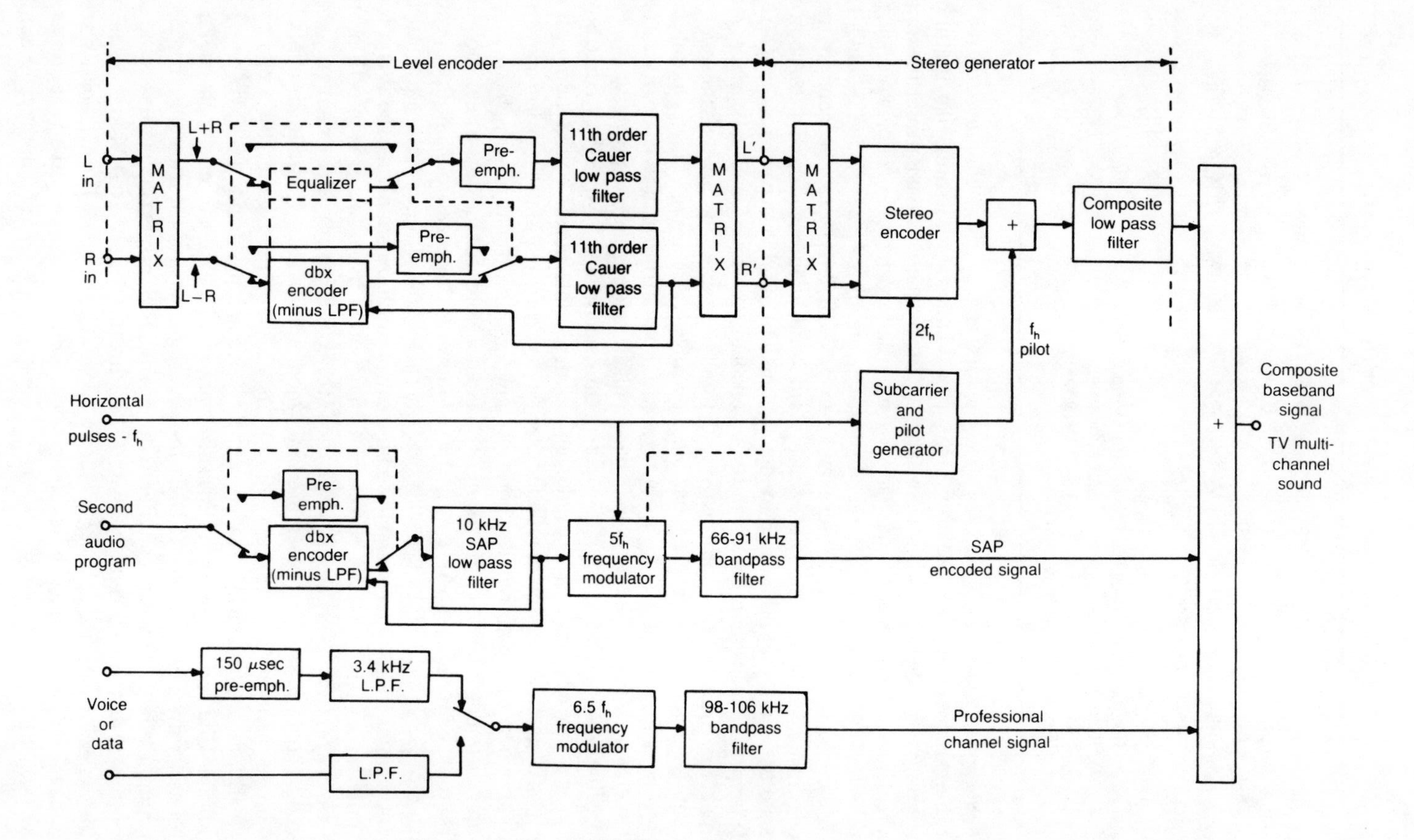

Fig. 9-2. Television multichannel sound—composite baseband signal (courtesy Zenith Electronics).

After filtering, the L−R and L+R are matrixed and passed on to the stereo encoder with its subcarrier pilots and composite low pass filter for composite baseband output.

dbx Encoder. Since both SAP and stereo (main channel) have the benefit of a dbx encoder, its outline and functional description may be conveniently undertaken here.

Figure 9-3 shows a block diagram of the dbx encoder as offered by Zenith. Prior to the Cauer lowpass filter, dbx appears in the level encoder and at the entrance to the second audio program (SAP) (Fig. 9-2). As shown in the illustration, dbx consists of feedback from the level encoded signal via pairs of fixed gain amplifiers, gain and spectral control bandpass circuits, and detectors, to both the voltage controlled amplifier and its spectral compressor. Any encoder effects due to parasitic problems or bandlimiting is duplicated in the L+R path as an equalizer, containing the usual 75 μsec pre-emphasis. With the compressor switched off, it is replaced by another 75 μsec pre-emphasis network. Following proper matrixing in the stereo generator, accurate right and left signals are then restored to the composite baseband output.

In addition to the dbx, level encoder, and stereo generator in the diagram of Fig. 9-2 are voice or data professional channel inputs with respective 150 μsec pre-emphasis for voice and 3.4 kHz low pass filter, followed by the specified 6.5 f_h 102.3 kHz modulator and 98-106 kHz bandpass filter. As said before, there's *no* pre-emphasis on the data channel.

Multichannel Decoding

It may seem somewhat gratuitous to include Zenith's original test decoder diagram and explanation at this juncture, but certain electrical information contained could aid some engineer in devising or modifying a piece of receive/monitor equipment for benefit of ego or financial reward—often inseparable and usually related.

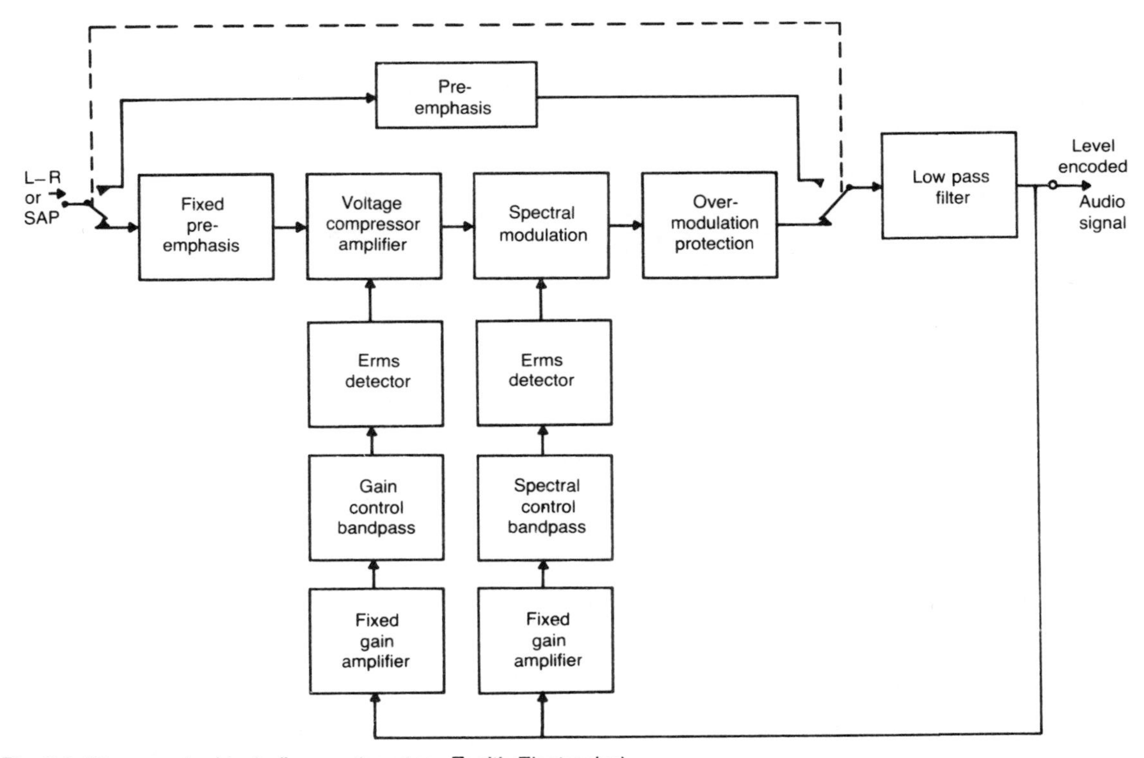

Fig. 9-3. Dbx encoder block diagram (courtesy Zenith Electronics).

In Fig. 9-4 is the original multichannel sound decoder diagram with filters and traps drawn, including component values. Using these as attenuation and pass filter guides, it shouldn't be too difficult to duplicate the original with, of course, a few well-chosen components and a good PC board.

At the inputs to this decoder are both a 5 f_h trap and second audio (SAP) bandpass filter. The first attenuates 5 f_h by 30 dB, and the SAP M-derived unit permits a 23 kHz bandwidth. The SAP lowpass filter has the same basic design as the cascade, M-derived π section 15 kHz filter with losspoles at 2 f_h and 5 f_h, but contains somewhat different components as you can see. SAP's bandpass filter losspole rests at 114.7 kHz, offering a symmetrical passband group delay.

Following the 4500 stereo decoder, the audio difference information contains undesirable f_h, 2 f_h and 5 f_h spurs and 2.5 f_h sidebands, necessitating subsequent filters. For best stereo separation and expansion a compensating circuit can be placed in the L+R section in cascade with 75 μsec de-

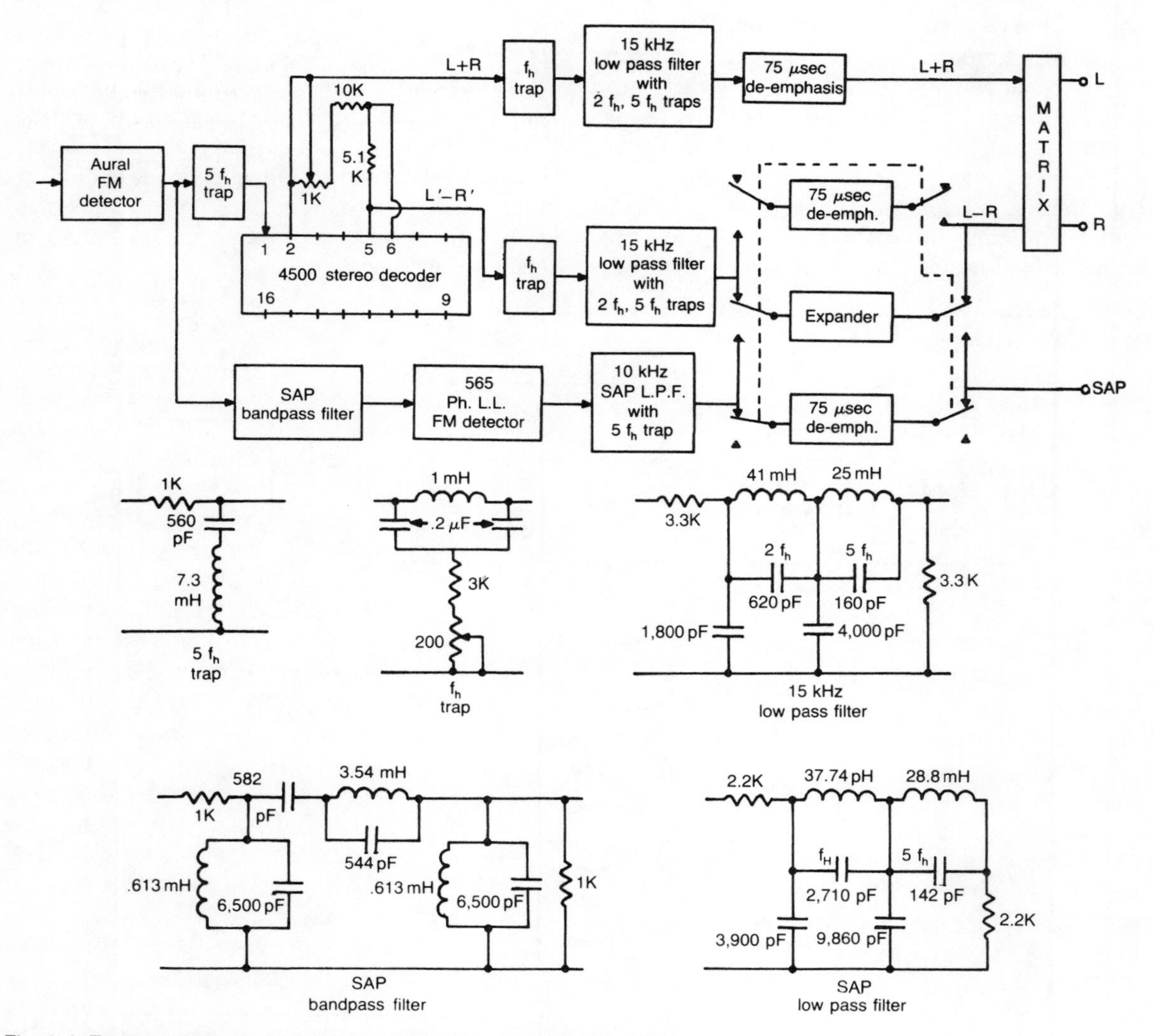

Fig. 9-4. Zenith's original test decoder (courtesy Zenith, Electronics).

emphasis, duplicating expander amplitude and phase response treatment resulting from parasitics and bandlimiting as in the encoder. As you see, both L+R and L−R have identical low-pass filters for best response and satisfactory separation.

SAP

Second audio program (SAP) has its bandpass filter followed by a 565-type phase-locked loop IC FM detector. A 10 kHz low pass filter with a 5 f_h series tuned trap follows to attenuate spurs above 10 kHz and the de-emphasis circuits block other spurious undesirables. Between the stereo and SAP signal paths is a single, switch-activated expander, and whichever signal isn't the expanded uses the 75 μsec de-emphasis circuit instead. In station monitoring, you may want to use two of these instead of just the one for more convenient readouts.

Although the FCC no longer requires specific monitoring, any broadcaster should know what his transmissions are doing. Zenith therefore suggests station operators keep tabs on the following:

Audio frequency response
Channel separation
Noise floors
Total harmonic distortion
Pilot/interface ratio
Signal/buzz ratio
Stereo subcarrier suppression
Crosstalk into stereo
Aural carrier peak deviation
(by all carriers)
Visual carrier IPM
Compression parameters

Carrier deviation peak flashers are also convenient.

The remainder of this chapter will be devoted exclusively to stereo TV exciters, monitors, modulators, and whatever else is available on the subject from broadcast manufacturers. It is interesting to note that more and more of these companies are rapidly awakening to the need and potential of this market and are actively planning and designing components to fill specific requirements. All have been contacted, and whatever they have seen fit to release will be included here.

TV TRANSMITTERS AND EXCITERS

Some portions of this section may neither be fully inclusive nor totally specific. Some will; some won't, due to intense design competition and a handful of electronic problems plaguing the industry ranging from dbx to stereo audio compressors and antiquated, narrowband transmitters. Most broadcast aural TV transmitter exciters are categorized as "indirect" FM, meaning a low frequency crystal and triplers to multiply the signal to whatever frequency you wish. These aren't especially linear and can't successfully handle stereo, direct FM exciters are required. Intercarrier phase modulation is another difficulty. Then there's broadbanding, and the addition of pre-distortion techniques to be considered, plus the problem of the notched diplexer, where many of these may have to be replaced. And there could be extra problems with retrofits for the professional channel's extended range at 6.5 f_h for audio processing and even 4800 bit/second data processing. So the path to good stereo sound, SAP, and the professional channel(s) hasn't been fully cleared by a long shot, and first class results will take some ingenious and studious doing.

As suggested in the beginning, TV multichannel sound broadcasting is due for a relatively slow startup—as late as the fall of 1985—but gaining momentum with a rush as transmitting and receiving equipment becomes available (with improvements) and public demand requires broadcast changeovers. Then as stereo TV and second audio programs become more commonplace, costs will tumble, designers accumulate initial systems experience, and quality sound for television could generate another round of receiver buying that should positively top all previous records. This significant development may also encourage better musical offerings from local and network stations who have been featuring contemporary "music" for tinsel ears. Perhaps the "big band" sounds of 1936-1948 could reappear . . . who knows? At any rate, stereo TV sound is certain to bring better audio to

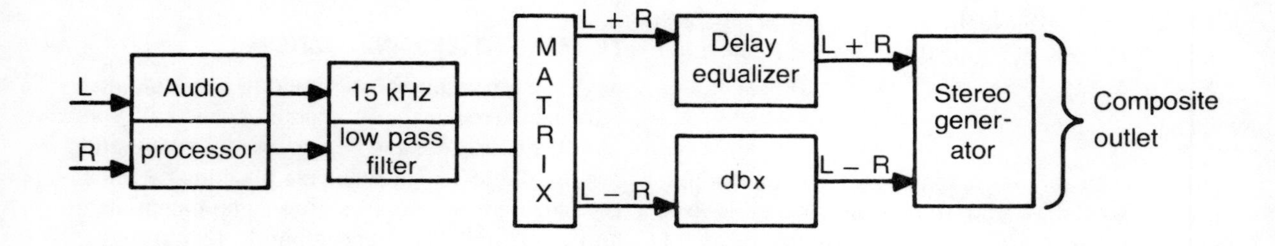

Fig. 9-5. Simplified block diagram of Modulation Sciences stereo exciter system. Audio processors supply compression limiting and threshold loudness. The 15 kHz filters are active rather than passive and are 70 dB down at 15,734 Hz, the TV line scan frequency.

America of one description or another and probably remain a part of consumer entertainment forever.

Modulation Sciences' Exciter (Generator)

A block diagram of Modulation Sciences' stereo TV system is ilustrated in Fig. 9-5. Left and right signals enter the twin audio processors and pass through 15 kHz low pass filters to the combining matrix. Here L+R and L− R are developed, with both delivered to the stereo generator and combined as a composite output. That's about the sum and substance of our block diagram for this portion of Modulation Sciences' release at this time. Fortunately, however, we do know what the system is intended to do, but not necessarily in logical sequence. In consequence, all functions will be listed as given by MS (metering functions are shown in Fig. 9-6).

☐ Mandated BTSC composite baseband.

☐ Audio processing—gated compressors and limiter.

☐ Semi-automatic loudness control.

☐ Optional auto loudness control.

☐ Built-in metering, avoiding the need for an external modulating monitor.

☐ Built-in stereo correlation (L and R) meter and out-of-phase alarm.

☐ Composite baseband output has sufficient drive for 75-ohm terminated cable, plus an optional full differential output for 3000 feet of cable.

☐ Unit requires only composite video input for sync lock.

☐ With video loss, unit automatically switches to mono and alarms.

☐ Audio inputs are left and right or sum and difference.

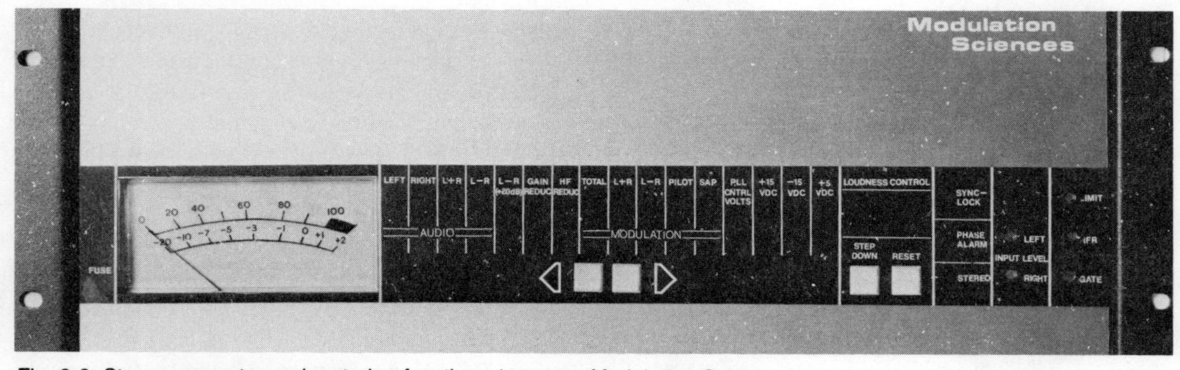

Fig. 9-6. Stereo generator and metering functions (courtesy Modulation Sciences).

☐ Baseband synthesized digitally by high resolution D/A converters.

Engineering vice president Eric Small emphasizes that his is the *only* company that's directly licensed by dbx to manufacturer its own dbx encoders—and there're plenty of ICs for hundreds of systems. The entire dbx, by the way, requires four ICs, one set for broadband and one for high frequency, and are surrounded by components with tolerances as close as 1/10 of 1 percent. The two sets of ICs work continuously and are placed in series. Other companies purchase dbx boards for the going price of $300 each.

Modulation Sciences SCA/SAP Generator

Change Subsidiary Communications Authorization (SCA) to Separate Audio Program (SAP) and you have at least the language translation between FM radio auxiliary services and those for TV multichannel sound. According to Modulation Sciences' V.P. Eric Small, the electronic changeover isn't that extensive and their new *TV Sidekick* (as can be seen from Fig. 9-7) is both in production and on 72 hours burn-in before shipment. Its measurements are 3.5H × 19W × 9.25 inches deep. Guaranteed temperature range extends from 0°-50° C, and all inputs and outputs are rf suppressed and power supply shielded. Signal connections mate through BNC plugs, jacks, and two sets of screw terminals. Frequency response extends from 50 Hz to 10 kHz, as BTSC specifies, along with dbx, Inc. noise reduction.

Using composite video inputs, Sidekick's subcarrier "unconditionally" locks to horizontal sync. Built-in dbx noise reduction results in signal-to-noise ratios of greater than 65 dB. The modulation monitor is a peak-reading deviation meter and there's an internal integrated audio processor. Significant audio gain permits telephone line direct connection. The synthesized subcarrier generator may be frequency programmed for other applications, and crystal-governed h/v pulses automatically assume control if and when there's TV sync loss.

A composite block diagram of this new TV Sidekick generator is illustrated in Fig. 9-8. Here you see rf and audio inputs with their synchronous measurement and input level controls, followed, respectively, by a measurement meter, filters, compressor, limiters and threshold control. The sync stripper and patch-through video permits sync pulse availability for frequency control for both the modulator and noise generator, with selected noise floor level adjustments. A harmonic filter receiving signals from the broadband limiter connects directly to the dbx encoder, with outputs going both to the modulator and the deviation position of the meter. The modulator then supplies its information to the mute circuit which may remotely controlled or defeated. SAP and compressor amplifiers then offer their separate outputs, including a connection for composite input.

Modulation Sciences says its TV Sidekick offers two methods of exciter connection, permitting "easy integration" for stereo TV audio with different types of aural exciters. In the loop-thru mode, composite baseband is buffered and mixed with SAP, and the combined output drives the wideband input of an aural exciter. Otherwise, a separate high level SAP signal may directly supply inputs to SCA circuits of many aural exciters. Sidekick's audio processor is *not* a standard FM processor, but has been specifically designed for subcarrier modulation that offers best voice and music reproduction with optimum S/N ratio. Ad-

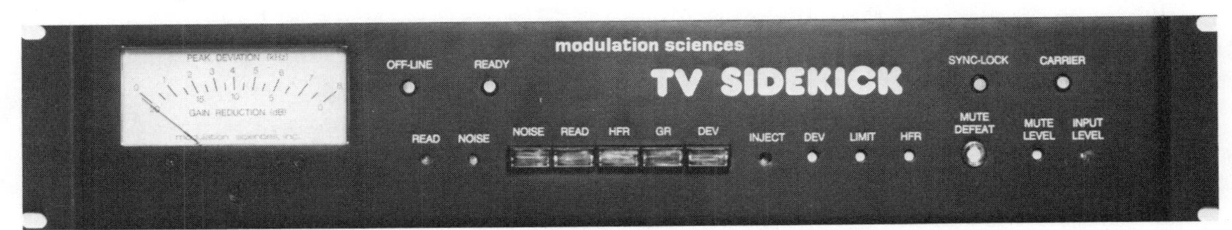

Fig. 9-7. Modulation Sciences' SAP generator and evaluator.

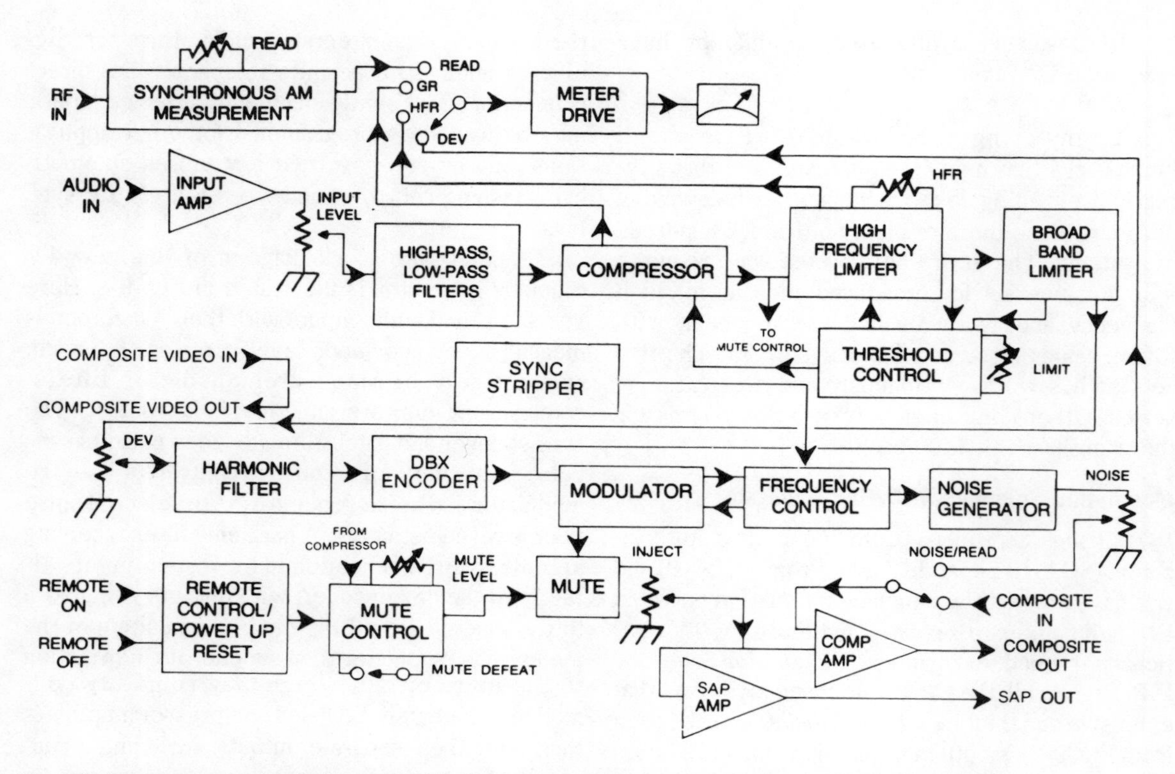

Fig. 9-8. Comprehensive block diagram of TV Sidekick SAP generator (courtesy Modulation Sciences).

justable broadband and high frequency limiter controls and compressor operating points permit customized audio processor response.

THE McMARTIN SYSTEM

Since Mr. McMartin refused either block diagrams or schematics, referring any such release to Zenith, we can only quote some manufacturers specifications which may at least be helpful to some extent. This was the system, of course, that did, indeed, use Zenith's encoder and decoder electronics for stereo, SAP, etc., and the company apparently cooperated completely with Zenith engineers. If there haven't been any substantial changes since BTSC testing, then we've actually covered the entire system several times over in preceding chapters. Therefore, I doubt if little is lost between theoretical and practical translations. If there are changes, those will show immediately on equipment schematics and probably written explanations also. So consider McMartin to be Zenith, and vice versa.

TV/Stereo Generator (BFM-7200)

This unit generates the 15.734 kHz pilot tone and doubles it for the stereo double sideband suppressed 31.468 kHz subcarrier. A summing amplifier to combine SAP and NP subcarrier feeds with composite stereo is included. You will also find disable switches for pilot, level encode, cross-couple, and the various L and R modes. Here are some parameters:

Frequency response—30 Hz to 15 kHz
Harmonic response—0.5 dB or less
Stereo separation—40 dB or greater
Crosstalk—40 dB
Subcarrier suppression—60 dB or better
L & R AF input Z—600 ohms balanced
Output Z—less than 50 ohms
Operating temp.—−20° to +50°C

Front Panel Controls

Switches	*Adjustments*
Pilot on/off	Pilot phase
Stereo/mono	Pilot level
Power	Comp level

Rear Panel Controls

Switches	*Connectors*
Cross couple	Composite out
Level encode	NP channel
	Sync in

Dimensions: 1¾ H × 19 W × 11 D

TV/SAP Generator (BFM-7000)

This unit offers an "ultra stable" vco which is phase-locked to the horizontal sweep, along with 78.67 kHz bandpass filter for SAP programs. There's also electronic muting, muting delays from 0.2 to 2 seconds, and modulation level adjusts from 3 percent to 100 percent. Carrier stability (free running) is specified at 0.25 percent.

Remaining principal parameters are as follows:

AF response—30 Hz to 10 kHz, ±0.5 dB

Distortion—<0.5% @ 400 Hz ± 10 kHz deviation

AF input—+10, ±2 dBm

Output level—0 - 6V p-p adjustable

S/N—65 dB or greater

Opeating Temp. —−20° to +50°C

Front Panel Conrols

Mute switch

Power on

LED carrier on

Frequency Adj.

Injection level

Rear Panel Controls

Level encode

Pre-emphasis

(in addition to

SAP output, input

sync input)

Dimensions: 1¾ H × 19 W × 11 D.

TV Baseband Aural Modulation Monitor (TBM-7500)

When used with the TBM-7200, this unit monitors all parameters of stereo transmission and, with TBM-7000, all parameters of SAP. By itself, the TBM-7500 monitors TV aural main channel modulation in broadcast stations. Total positive and negative modulation is metered, along with S/N as low as −70 dB, while a peak flasher will indicate highest positive or negative peaks. Thresholds may be adjusted between 50 and 120 percent.

The meter acts as a semi-peak reading voltmeter for modulation, and is damped for AM/FM noise measurements, typically −75 dB below 100 percent modulation. Low and high level cards are optionally available. When rf fails or falls below a preset level, the front carrier lamp goes out and audio output is automatically muted. There's a 600-ohm balanced output for audio monitoring. Operating and modulation ranges are specified as 54-212 MHz (4.5 MHz intercarrier) and 73 kHz deviation at 100 percent modulation. Standard rf inputs are at 50 ohms with sensitivity between 0.1 and 1 watt.

There are outputs for audio monitoring, distortion, and composite. Audio frequency response is from 30 to 15 kHz and composite from 30 to 120 kHz. Dimensions are: 5¼ H × 19 W × 13 D.

ORBAN ASSOCIATES SYSTEM

About all the information we have on this very new TV sound transmission system is a single page of description and an accompanying simplified block diagram (Fig. 9-9). At this stage of general multichannel equipment, we're probably fortunate to have that much since it has yet to become what's known as "public information." Unfortunately, there isn't even a spec sheet detailing any of the parameters. All we have is included, nonetheless, in order that maximum knowledge may be offered even though some portions are obviously incomplete. But we do thank Mr. Orban for his consideration in making even this much available.

As Bob Orban states, there are four basic sections to this block diagram that permit introduction of right and left stereo inputs, the second audio

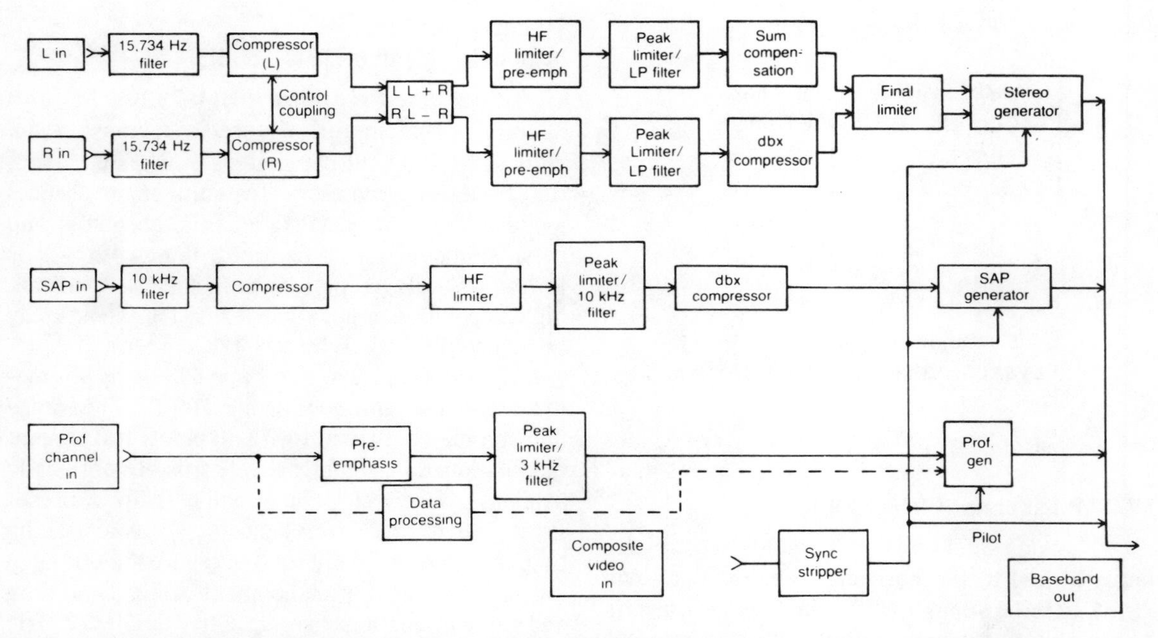

Fig. 9-9. Preliminary and simplified block diagram of Orban BTSC generator (courtesy Orban Associates).

processing channel (SAP), and the professional channel. After the 15,734 Hz filters, both right and left inputs are compressed to some degree by "control coupling," procedures and delivered to a matrix which generates the sum and difference components. L+R is then high frequency-limited, pre-emphasized, peak limited by passing through a low pass filter, compensated, limited once again along with the stereo channel output and delivered to the stereo generator.

The L−R signal is both high frequency limited and pre-emphasized, then peak limited and low pass filtered before reaching the dbx compressor. After that, final limiting (with L+R) takes place and the stereo generator output is applied to the transmitter, picking up pilot signals along the way. It's synchronized by sync-stripped video which also generates the pilot and 2 f_h stereo subcarrier.

The SAP second audio program begins with a 10 kHz bandpass filter, is then compressed, high frequency limited, 10 kHz peak limited, and channeled into a separate dbx compressor for eventual SAP generation. Peak limiter output goes to an FM subcarrier generator, phase-locked to 5 times f_h and then into a phase-linear bandpass filter, preventing both "out-of-band emissions and interference to the stereo subcarrier from SAP."

The *professional channel* is narrowband voice/data with carrier at 6.5 f_h. There's pre-emphasis for voice, peak limiting and a 3 kHz filter, all of which is bypassed in the data processing mode. The FM subcarrier is phase-locked to video.

Chapter 10

Multichannel Sound TV Receivers

Strange as it may seem, tackling this final chapter is a real challenge but not a chore. In the first go-round on any new product, all cards are held tightly "to the vest" until competition rises to the challenge. Here—although the results of multichannel sound should appear better than average—the rapture of complete satisfaction probably won't surface for some time to come. True, you'll have stereo-ready and stereo equipped receivers on the market as 1984 turns into 1985, but broadcast stereo exciters and dbx companding both are having growing pains momentarily, and TV transmitter modifications may require somewhat more work than originally expected to place a number of broadcasters on the air, let alone the three or four major networks.

In the interim, the networks will do special programming in TV audio stereo, some of your local stations will also be on the air with particular events, and a few public broadcast stations may already have accessible equipment to develop a head start. But as 1986 arrives with several of the TV networks substantially or completely converted to satellite transceiving, a great deal of stereo and SAP (separate audio programming) should be available to all those with receivers so endowed. After that, receiver and transmitter refinements could come rapidly. Eventually, satellite-transpondered multichannel sound, picked up and broadcast by improved terrestrial transmitters, should present interesting and gratifying results indeed. Shortly, thereafter, the U.S. may become, at last, partially a bilingual nation, with English and Spanish predominating, followed by others with city or regional influences. At least, that's the outlook as pieced together from broadcasters and knowledgeable industry sources who mostly wish to remain anonymous.

THREE TYPES OF RECEIVERS

There are three preferred types of multichannel sound receivers, at least two of which you're going to see in 1985-1986 production. (Fig. 10-1). The first would entail completely separate tuners for separate sound and video processing. The second has a single tuner, but would produce entirely isolated video and sound channels for amplification

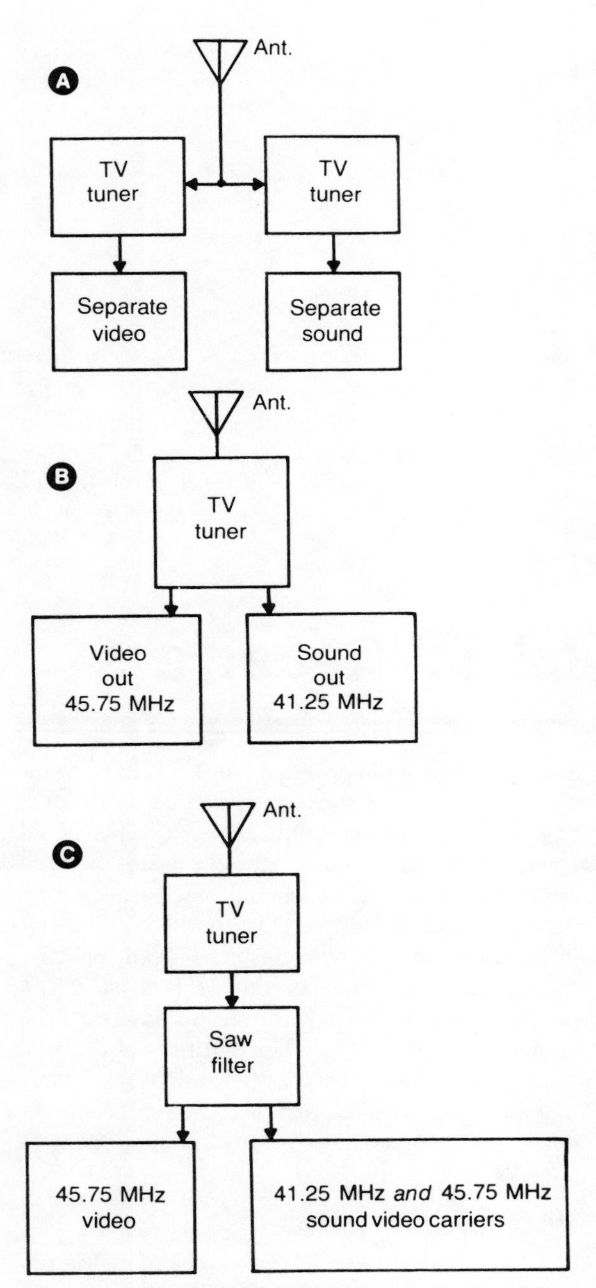

Fig. 10-1. The three possibilities for stereo TV sound: (A) separate sound and video, (B) split sound and video, (C) quasi-split sound and video.

and eventual video/audio detection. The third—and most popular here and abroad is the quasisplit sound which divides the aural and visual signals only after the surface wave acoustical (SAW) filter, following a single, unified tuner. Each of these systems has advantages and disadvantages as we shall see. Other already introduced systems differ.

Thanks are due to Pieter Fockens and Victor Mycynek of Zenith Electronics Corp. who supplied, through written and oral aid, the ability to understand and then record what could be to many, a highly complex audio processing system. Charles Smaltz of National Semiconductor is also due equal credit for integrated circuit information, without which this chapter would supply little more than a theoretical overview. So to Pieter Fockens and Charlie Smaltz, a multitude of thanks for lots of valuable information and illustrative diagrams for which they receive due credit. Illustrations without credit lines were modified or originated by me.

Split sound and separate sound receivers are normally impractical due to buzz problems and cost. Since the separate sound sets always require individual tuners, not only would the package be bulky, but another $100 or so might be added to already expensive merchandise and sales could suffer. Split-sound receivers with a single tuner, however, are subject to thermal noise and intercarrier buzz problems and are not suitable for cable TV. Conversely, quasi-split or quasi-parallel intermediate frequency amplifiers with separate detectors can eliminate buzz by a refined 4.5 MHz intercarrier process and also reduce or remove incidental carrier phase modulation by Nyquist slope rejection. Therefore, many popular makes of receivers with "stereo ready" plugs will appear as quasi-split/parallel (same thing) types, and the appropriate adapter may be attached. Some others will be designed for SAP (second audio programs) only, and their output jacks may be unmarked or simply emerge as an RCA-type phone plug.

After the IFs, of course, there are the 4.5 MHz intercarrier detector, stereo and dbx decoders and ultimate L and R matrix for twin amplifier-speaker connections. The EIA-BTSC (Electronic Industries—Broadcast Television Systems Committee) decided the composite baseband signal requires some 90 kHz and so an input of 180 kHz is needed at the bandpass filter of the 4.5 MHz intercarrier. Either doubly tuned circuits or ceramic filters offer good bandwidth and sufficient selectivity, although the FM detector response has to be widened and linearized. According to Mr. Fockens, this prevents intermodulation products from causing SAP-stereo crosstalk. Further, because the L+R mono signal is not companded, better S/N ratios are needed in the detector since left and right audio noise and buzz will be controlled to a great extent by these electronics—something that the industry is already working on. It has been solved to some additional extent by dynamic noise reduction integrated circuits now supplied by National Semiconductor.

Just before final manuscript editing in preparation for printing, we also discovered that one or more offshore manufacturers will attempt to use the *usual* 4.5 MHz intercarrier takeoff normal to non-stereo sets, demodulate it, and then route composite baseband audio (with pilot, $2f_h$ and $5f_h$ carriers) directly into the BTSC-dbx decoder. This method is certainly cheaper but unless the Japanese know something we don't, it could create problems.

Now, before we launch into the actual detection systems and a dbx companding and expanding, let's do a description of a very useful IF integrated circuit that can be used at least in the video portion of stereo/SAP receivers with good results.

The LM1822 IF

A very good quality video intermediate frequency amplifier with 4.5 MHz intercarrier audio detector and synchronous video detection, National Semiconductor's LM1822 is not the only IC of this description available but is being used by a major TV manufacturer with considerable success. Suitable for both home receiver and cable IF applications, it will accommodate all video rf/mixer inputs from 38.9, 45.75 (usual), 58.75, and 61.25 MHz at selected points on the input response. The 28-pin, molded dual in-line package (N) features true synchronous video detection (with PLL), noise-averaged gated AGC, AFC detector with adjustable bias, NPN video amplifiers, white noise inversion, adjustable zero carrier level, and good stability at high system gains. Chip power dissipation measures 2W max. at 15 Vdc, with IF gain usually 35 dB. A typical application of the LM1822 is shown in Fig. 10-2.

As you see, the surface acoustical wave (SAW) filter has a single tuner input and two outputs into the 4-stage intermediate frequency amplifier. This is used to reject unwanted rf coupling problems via differential inputs. IF gains are adjusted by various values of an external (and fixed internal) resistors at pins 3 and 4. Pins 7 and 8 permit dual SAW filter inputs to the emitters of common base amplifiers, while pins 6 and 9 decouple the dc feedback loop from respective bases. Common base amplifier gains depend inversely on source impedances which may vary between 500 and 2,000 ohms in SAW filters.

Pin 11 connects to an open-collector NPN transistor that begins to conduct when voltage at pin 13 exceeds the fixed voltage set at pin 12 by 0.6V, and becomes reverse AGC. At the same time, preset voltage at pin 12 determines when the IF quits gain reduction and the tuner begins gain reduction.

The video detector and limiter receives outputs from the IF amplifier as additional inputs from the voltage-controlled oscillator and variable phase shifter contribute to linear demodulation. There is a dc phase adjust potentiometer as well as an external resistive divider to set the gain of the phase detector at vco control center. For maximum efficiency, the vco needs a signal reference precisely in phase with the carrier input, but PLL operation sets vco phase in quadrature. Therefore, a variable phase-shift network between the vco and video detector must compensate with proper phasing as current amplifiers and synchronous switches undertake their job of baseband detection.

A parallel LC tank circuit centers the AFC (auto. frequency control) for tuner channel lock between pins 23 and 26, and at the same time another LC tank at terminals 24-25 contributes the tuned load for a single stage limiting amplifier, stripping amplitude excursions from AFC and phase detectors. AFC detector output becomes a push-pull current, single-ended output, with an AFC defeat

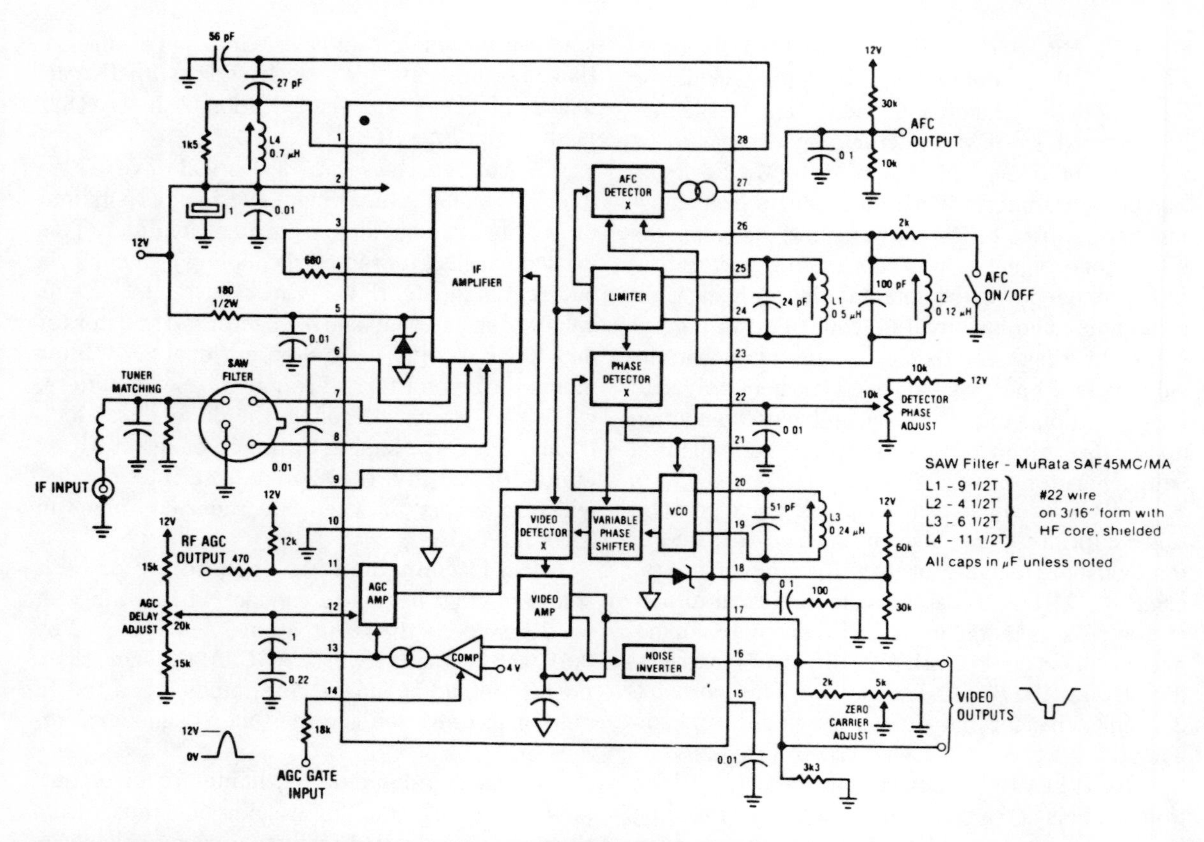

Fig. 10-2. National's audio/video processor IF chip (courtesy National Semiconductor).

switch across the AFC centering tank, if desired.

Video output is negative-going to place noise in the normally unseen black picture area. But with no detector input, noise inverter output rests at about 0 voltage, and usually decreases as input information increases. When noise causes the PLL output to rise above zero carrier, the noise inverter inverts these resulting white pulses, making them gray and less noticeable.

Above pin 16 is pin 17, where there is a no noise inverter and conventional sound takeoff occurs. Here, noise becomes "averaged out" by a low pass filter which also contributes to detector linearity whenever there's overmodulation, permitting "excellent" sound subcarrier linearity. Negative sync tips are clamped at pin 17 to 4 volts and the zero carrier level may be adjusted by an internal 500-ohm resistor in series with the low impedance emitter output by changing the load current. Gain paths of the video amplifier to pins 16 and 17 are matched so that the two outputs track.

For the sound section—if you're really doing quasi-split sound—another IC will be required to process this second portion of the dual IF train and individual detectors. An LM1823 can be used, according to National, or some other appropriate IC that's compatible with the '22. From what appears in succeeding information, most of the receiver manufacturers have their own ideas and are using whatever types they wish. You'll see this come to pass as we begin to review the actual receiver systems delivered to us by the major manufacturers. Much of the material is totally new and being generated and published for the first time. Nonetheless, you'll find considerable coverage among the major multichannel systems, which should stand you in good stead for a number of years to come, with the usual annual refinements to be added by those who maintain contemporary interests.

THE FINAL TYPE

The LM1822, naturally, can be used with almost any good tuner and appropriate dc power supplies. But for stereo/SAP processing, this is only the beginning, and the basic block diagram of a quasi-split sound receiver is shown in some detail in Fig. 10-3. When 41.25 MHz audio and 45.75 MHz video are beat together, the difference becomes 4.5 MHz intercarrier sound and is ready for further filtering, FM detection, and decoding for SAP/stereo-applications as shown. The process appears somewhat complex—and it is—therefore the best way to tackle this receiver portion is to describe the action with available building blocks so that when receiver diagrams are available you'll understand the basics well enough to "plow" right through them.

Stereo Decoder

The original stereo decoder turns out to be a 4500 integrated circuit, made by several companies and identified as an LM4500A by National Semiconductor and a TCA4500A by Motorola. This unit has been valuable in eliminating second and third harmonics of L– R, keeping SAP/crosstalk out of stereo. But, as Pieter Fockens says, there is need for a new IC. One that: offers pilot detector immunity from SAP; a narrower acquisition range for pilot PLL; separate L– R and L+R outputs; and elimination of pilot from L– R and L+R outputs, saving more traps. A second generation 4500 IC type, the LM1884, Fig. 10-4, already does some of these things and is another step toward quality TV stereo sound. It permits compressed L– R stereo extraction for dbx expansion. At the same time, restored L– R is matrixed with L+R elsewhere and left and right audio becomes available to switching, tone and volume circuits.

National's LM1884 is a specific decoder especially designed for TV stereo. It features low impedance L+R outputs, a mono/stereo switch and indicator, typically 0.1 percent distortion, buffer amplifiers, pilot detector, oscillator, dividers, decoder, and mono and stereo outputs. With the oscillator you'll also see a voltage-to-current amplifier and PLL phase detector (Fig. 10-4). Gain ratio between L+R and L– R ranges between –2 and +2 dB, with dc mono to stereo output shift ±20 mV max. At 25 mV rms for pilot, the minimum capture range is ±0.5 percent, and less than 20 millivolts turns the pilot lamp on and off. Supply current for this chip at 16 volts Vcc measures between 15 and 50 mA, with the signal output imped

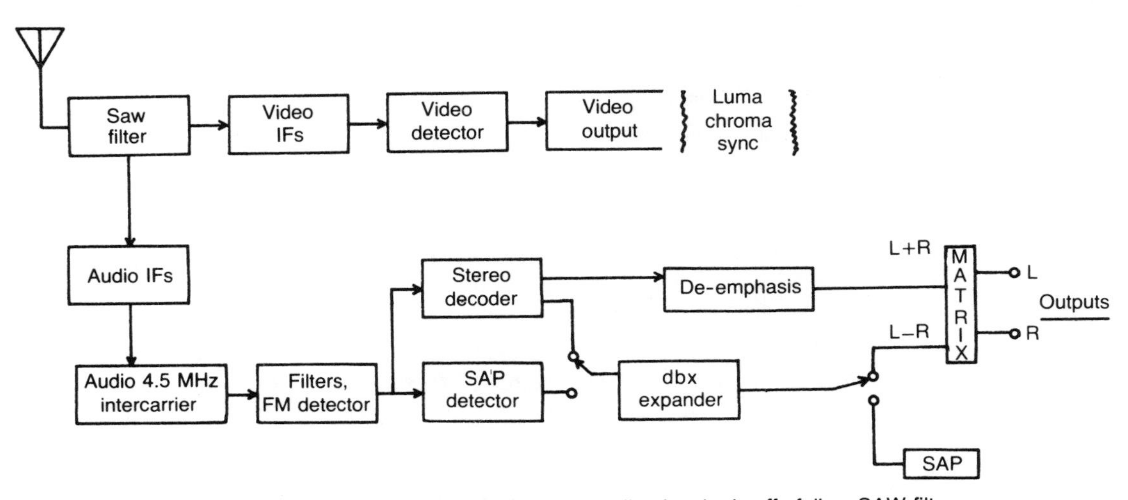

Fig. 10-3. Basic block diagram of a quasi-split sound receiver. Audio-visual takeoffs follow SAW filter.

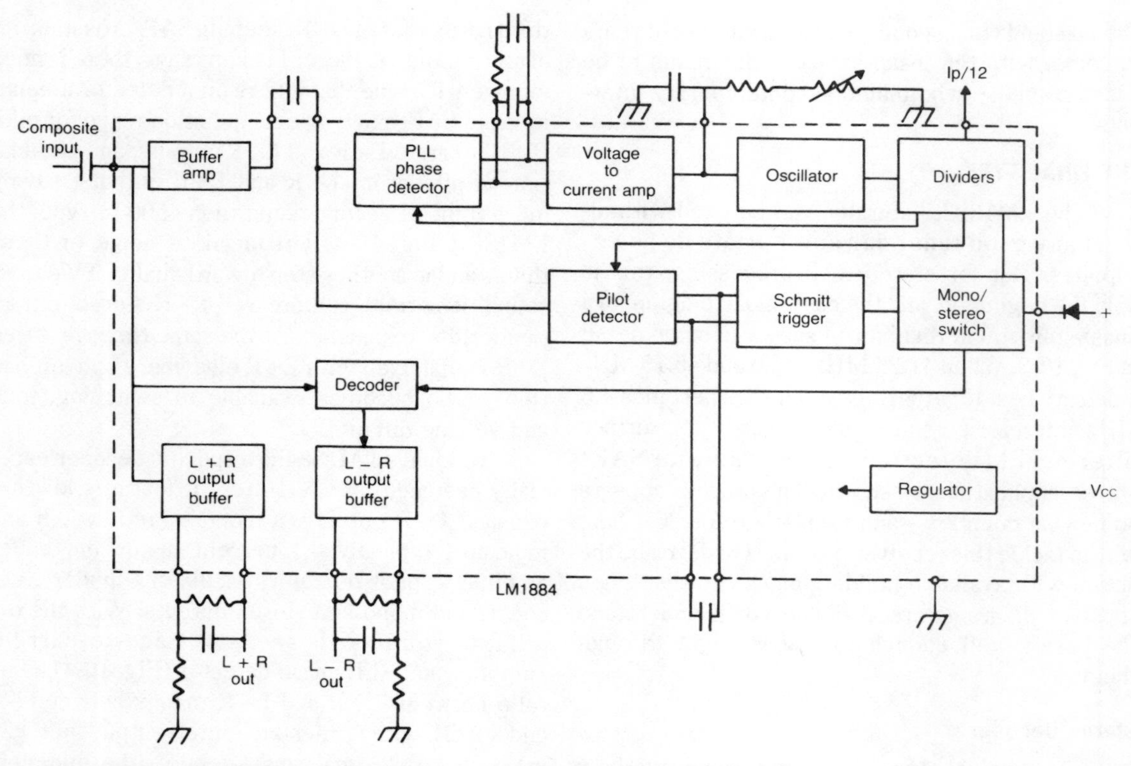

Fig. 10-4. National's sophisticated TV stereo decoder (courtesy National Semiconductor).

ance typically 100 ohms.

When the pilot tone is detected and monophonic switches to stereo, the decoder supplies L−R to the output buffer which operates simultaneously with the mono signal L+R buffer, producing full stereo for final L and R matrixing in another IC. The diode on the right of the diagram represents the stereo LED indicator. There's also a manual stereo/mono switch and a potentiometer vco outboard adjust connected to the active circuit.

SAP Detector

The separate audio program (SAP) is not a high quality audio channel since max. modulating frequency is only 10 kHz, and peak deviation but 15 kHz (compared to even present mono. that has 25 kHz deviation). Therefore, conventional LM565 FM detection is quite adequate for the purpose. A SAP bandpass filter (not shown) precedes the SAP IF, limiter, and PLL FM detector. Fortunately, at a frequency of 5 × 15,734 Hz horizontal scan, very little filtering is needed between SAP and L−R, and a bandpass filter, according to Mr. Fockens, will avoid capture effects. But a narrow passband is said to increase distortion before dbx companding (expanding in the receiver).

SAP may be detected by either a pulse counter or by PLL used as a discriminator, such as done in the 565 IC. It has high loop gain, good hold-in range and high deviation-to-carrier frequency ratio. Mr. Fockens says a carrier threshold, however, is needed in the SAP channel to either mute the channel when there is no SAP, switch to mono or stereo, or protect the channel from other SAP transmissions. PLL designs for stereo pilot and SAP detection will follow sudden carrier phase changes and maintain lock so that transmitter sound and picture switching can proceed routinely.

However, before we begin to describe the

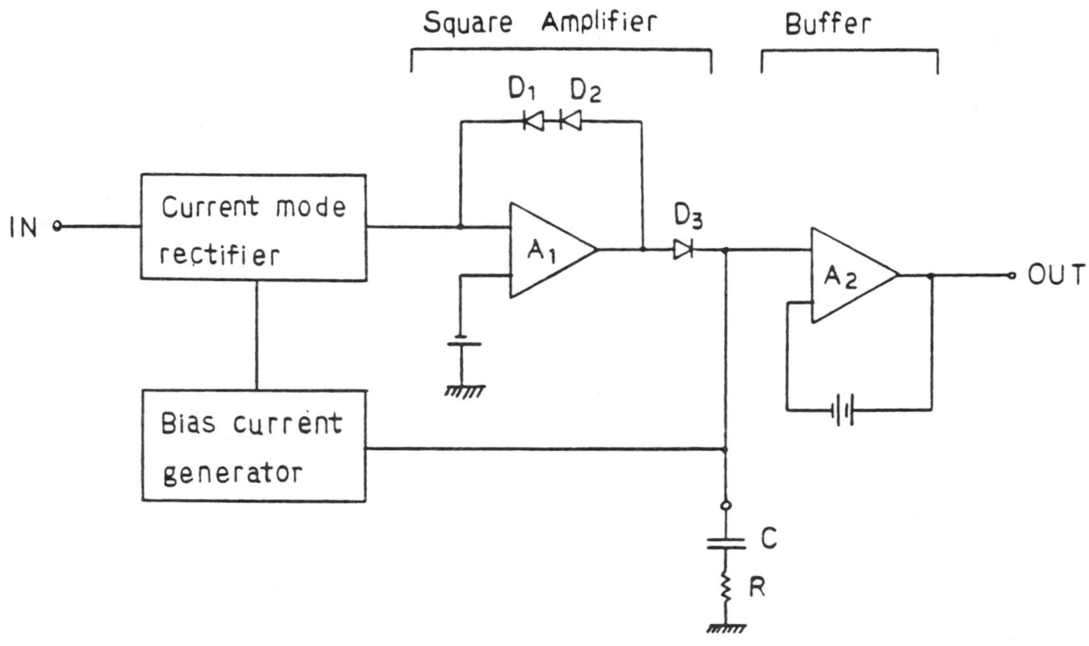

Fig. 10-5. Block diagram of dbx noise reduction system (courtesy NEC).

various multichannel TV sound systems specifically, it might not be a bad idea to review dbx once more—this time in considerable detail—so there'll be no misconceptions as to its purpose and operations. As previously stated, one of the prime hangups in the entire stereo/SAP application is the effect of the compander (expander) on the remainder of the system. If very close tolerance components are not used and the "humors" adjusted just right, you'll hear considerably more noise and distortion than is agreeable. That's the reason for our cautioning that some receiver systems will definitely sound better than others—at least in the beginning. After several years, "standard" ICs will have reached the market and almost any designer can fit them into his television receiver. Having said all that, let's go more directly to a very explicit discussion of dbx, detailing exactly what it does and why.

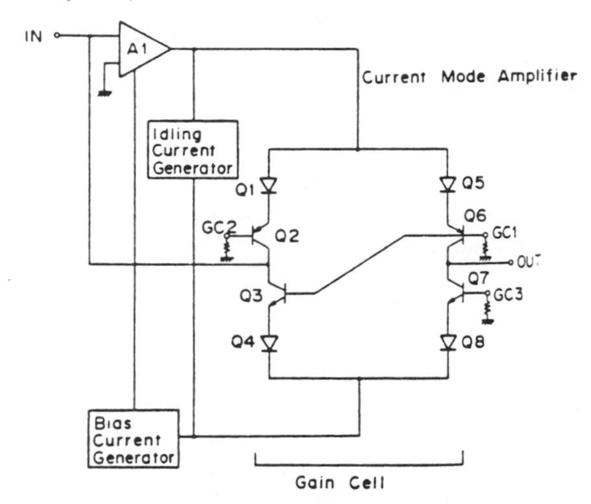

Fig. 10-6. Signal flow diagram of VCA IC (courtesy NEC).

The dbx Encoder/Decoder

A somewhat better understanding of the dbx system can be gained from a short discussion of a paper presented before the IEEE (Institute of Electrical and Electronic Engineers) Consumer Electronics group in late 1982 by Nippon Electric Co. engineers Koji Shinohara, Mamoru Fuse and Yuji Komatsu. With the aid of Dr. D. E. Blackmer of dbx, these three developed the voltage controlled amplifier (VCA) and root mean square value level

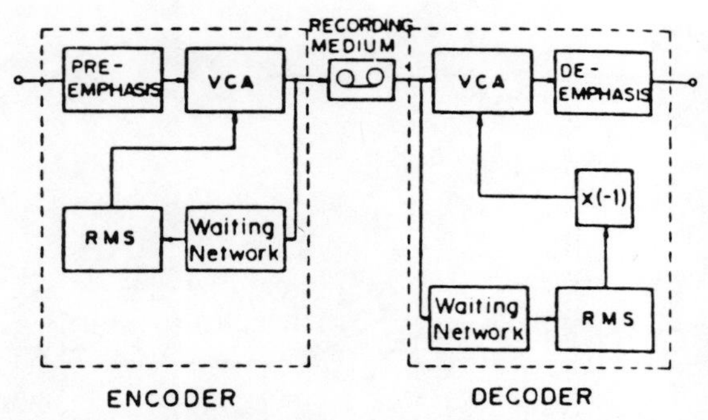

Fig. 10-7. Signal flow diagram of rms IC (courtesy NEC).

sensor (rms) integrated circuits for NEC and U.S. clients.

A simple block diagram of the encoder and decoder is illustrated in Fig. 10-5. Here are shown the familiar pre-emphasis, then voltage controlled amplifier, followed by a waiting network and the rms leveler feedback. Received information enters the VCA first, with the waiting network, which supplies the rms chip, as before, then passes through an X(−1) block before return to the voltage amplifier and the final de-emphasis stage.

In all of this, the VCA becomes a dynamic range compressor in the encoder and a range expander in the decoder. This occurs as the rms chip gain-controls VCA output proportionally to the root mean square value of the input. Linear expansion and compression amounts to a factor of two, with the original dynamic range (in dB) halved during compression and doubled in expansion—with recording levels advanced and noise levels reduced because of the resulting dynamic range.

A drawing of the VCA is shown in Fig. 10-6. Transistors Q1-Q8 constitute the gain cell, with the Q2 emitter signal a logarithmic function of the negative input current, and the Q3 emitter, a logarithmic signal proportional to the positive input. The overall input, of course, passes through current mode amplifier A1 to the gain cell and is, in turn, regulated by a bias current generator. The output between Q6 and Q7 consists of antilog of these developed signals to that I_o equals I_{in} and VCA gain develops antilog control by the control voltage itself.

Figure 10-7 illustrates the RMS integrated

Figure 10-7 illustrates the rms integrated circuit. Here, diodes D1-D2 shunting A1 form a log circuit with an output "proportional to the log of the absolute value of the input . . ." Diode D3 and the RC constitute the rms analog circuit supplying the antilog function and delivering an output voltage proportional to the rms input, which is amplified and buffered by A2. In reality, all this takes about 25 transistors and more than several diodes to accomplish what appears so simple in the basic block. The same is true for the VCA.

Therefore, considering the logarithmic expansion, contraction, noise reduction, and thermal stability of these two chips, such an undertaking is not nearly as simple as one of these abbreviated writeups would make it seem. Complex ICs are tough from many standpoints and their development and uses should be duly appreciated, especially when log and linear functions are combined to produce mixed and predictable results. It's not difficult to imagine that a great deal of design engineering went into this particular set of ICs (or Matsushita's single IC) to yield desirable companding products compatible with other ICs that carry on the linear, stereo, and SAP processing. Professional channels, by the way, are *not* companded.

CONTEMPORARY MULTICHANNEL SOUND RECEIVERS

As promised, we will review available multichannel receivers in this section, with particular emphasis on those now leading the industry. As you will observe, most have outboard attachments for stereo/SAP options, although RCA offers built-in, all-inclusive stereo/SAP in many of the more expensive models with no external fittings at all. We'll have to see how the two philosophies fare in the marketplace, although there's due suspicion everyone will thrive because of outright novelty and consumer demand. Later, when comparative products are freely available, selectivity will probably take over, just as it has in market response to quality color receivers; and as their dominance grows, everyone will have more advanced and better products. Just don't expect every system—whether receive or transmit—to be perfect. A few months following the FCC's Zenith-dbx decision really isn't enough time to produce perfection. Those with sufficient foresight and engineering talent, however, do and did produce a pretty reasonable product. But in 1985 and beyond, you be the judge.

RCA's Multichannel System

In addition to being the only maximum bandpass luminance and chroma receivers marketed in the United States (or probably anywhere), RCA's 2000 series will also feature 18 top-of-the-line models handsomely equipped with the latest and possibly the best in multichannel TV sound. For not only is RCA using entirely separate video and sound IFs *following* the input buffer, but very careful filtering before the 4.5 MHz sound detector and further sound trapping to the audio demodulator itself. Separate from the chassis and located on a 3-circuit board assembly, these assemblies consist of sound IFs, demodulators and matrix, the dbx expander, audio volume/tone controls, and audio outputs. Not especially identified but highly important is the Dynamic Noise Reduction IC (DNR) from National Semiconductor which should aid immeasurably in producing very efficient, high quality speech and music sounds, even though noise reduction is somewhat proportional to bandwidth restrictions during its operation.

The LM1894 is a single chip IC that will attenuate high frequency noise without producing noticeable changes in program material. With preset threshold, this IC is designed to pass full audio bandwidth information at maximum signal input, but introduce a 6 dB/octave low pass filter as inputs decrease toward this preset level. Con-

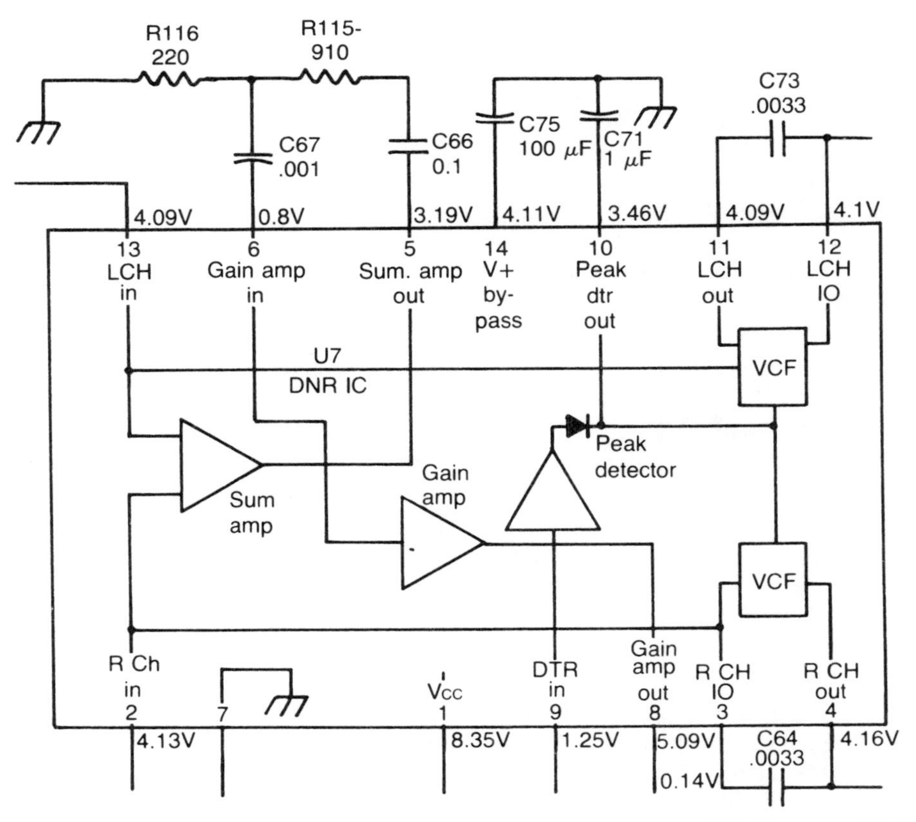

Fig. 10-8. Detailed block diagram of RCA's dynamic noise reduction circuit (courtesy RCA Consumer Electronics).

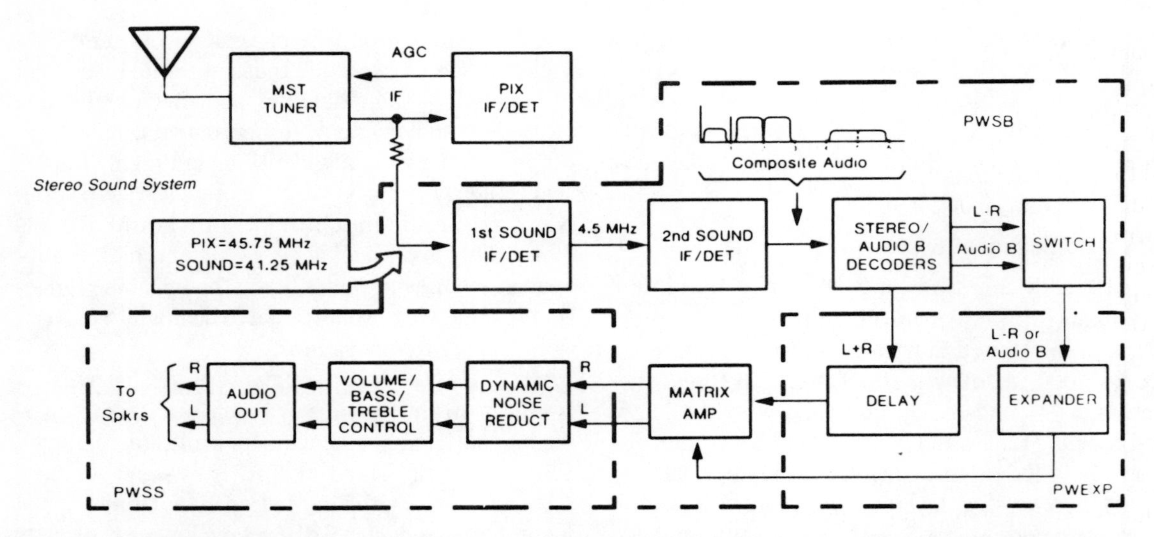

Fig. 10-9. Simplified signal flow block of RCA's stereo and SAP sound circuits (courtesy RCA Consumer Electronics).

versely, as audio signals increase, bandwidths increase toward −3dB at 30 kHz, permitting up to 14 dB of noise reduction. Under no signal conditions, mid- and high-frequency noise is attenuated. Applicable to tape players and TV audio, this IC may also be used with auto radios, hi-fi systems, and VCRs to effectively reduce both hiss and noise in programmed information. The DNR is normally located after audio detection and before volume, tone, and balance controls with optimal inputs of approximately 300 mV. To illustrate this IC, we'll show the simplified diagram RCA actually uses in its CTC 131/132 chassis as Fig. 10-8.

Indentified as the U7 DNR IC, this integrated circuit turns out to be a 14-pin in-line, plastic configuration, with right, left, and gain amplifier inputs, a summing amplifier, peak detector, and voltage-controlled filters. According to RCA's Technical Training manual, left and right audio signals are summed and then filtered after initial inputs, gain amplified, and passed through an external bandpass filter permitting low outputs between 20 and 1,000 Hz, with higher levels at about 7 kHz. After bandpassing, there's further amplification and then peak detection, producing a dc output proportional to input. Two voltage control amplifiers, connected both to the peak detector and the original right and left outputs, vary their individual bandwidths as peak detector levels rise and fall. At greater inputs, high level control voltages *increase* bandwidths as S/N ratios become less critical.

A block diagram of the entire sound system has also been generously supplied by RCA (Fig. 10-9). We'll use RCA's latest receiver schematic as well as further information supplied by that Company's able Consumer Products Technical Division. After 44 MHz center response inputs reach and pass the input IF buffer, a 45.75 MHz bandpass filter permits video into one end of the dual processing IF IC, while a 41.25 MHz bandpass filter admits full-bandwidth audio into the other. Here video is synchronously detected, AGC generated, and the 45.74 and 41.25 MHz carriers beat together to form a conventional 4.5 MHz difference output, avoiding the buzz-producing Nyquist slope problem as well as incidental carrier phase modulation and tuner spurs by 4.5 MHz intercarrier sound detection. Series resonant and parallel resonant passive LC elements then couple this audio to the sound limiter and demodulator with its outboard RLC adjustable tank circuit—called a quadrature detector.

Broadband audio becomes level adjusted and RC coupled through a bandpass reject amplifier and its emitter $1f_h$ trap, greatly attenuating the 15,734 BTSC-assigned stereo pilot frequency. Composite audio information, however, is capacitatively coupled to a switch IC which, if in the stereo position, returns these frequencies to the stereo decoder for processing. If in the SAP, or audio B position, restricted aural signals are processed through the $4f_h$ and $6f_h$ audio traps to the second audio program decoder which operates at the $5f_h$ carrier frequency of 78.67 kHz. An RC adjustable oscillator does the detecting. Should the stereo switch be placed in its monophonic position, the SAP decoder is cut off and its output muted. The two narrowband $4f_h$ and $6f_h$ traps on either side of the SAP carrier help attenuate any residual buzz.

The bandpass rejection amplifier inverts and amplifies the 15,734 Hz pilot signal and, whenever it is apparent, syncs the phase-locked loop (PLL). This frequency is multiplied by two for the stereo subcarrier; developing the left AM double-sideband suppressed information, paralleled by right mono. But since dbx had companded L−R, outboard transistors Q8 and Q13 must once more develop the true L−R and L+R.

Now L−R or SAP is switched into the dbx expander, L+R becomes delayed to match the propagation time of L−R or SAP, and both mono or stereo/FM now reach the matrix amplifier together. Beforehand, seven integrated circuits on the dbx expander filter, amplify, and rms detect L−R, passing one de-emphasized output to SAP and the other to audio control, along with phase-compared −(L+R) from the stereo demodulator. Right and left signals now reach their respective speakers. With stereo, VCRs, discs, digital hi-fi, radio, etc., remote controls are usually hard put to keep up with these various attachments, let alone control their operations from a distance. But something is already being done about this. In the U.S. and abroad, RCA and Philips seem to have a head start toward the day of the household home-entertainment console.

The Equipment Bus. Just as computers, certain test equipment, video terminals and the like are, say, 488 IEEE bus compatible, so must consumer products develop this capability also. For instead of multiple plug-ins into some black box with various amplifiers, losses, gains, and dubious impedances, video and audio products must soon conform to a workable standard, permitting acceptable, low-loss couplings for all inputs so that speakers and cathode ray tubes may successfully reproduce at baseband the incoming rf and audio/video information. With such a setup, it's then possible to program and select various inputs, allowing each to be visible or audible on some specific display for desirable convenience of any system operator.

In a session with design engineers at Indianapolis headquarters, RCA demonstrated just such a system they called "piggy back" (Fig. 10-10) with "daisy chain" connectors. It's really a 50-foot cable and 16 devices which will accommodate 13 separate equipments, including four of any one product such as video cassette recorders. The video portion is terminated in 75 ohms, while audio mates in the kilohms. With two microcomputers in the new 26-inch receiver, all commands from a multi-programmable remote console register on the TV screen for about four seconds. There is also a system status recall that can produce five pages of functional directions sequentially on the CRT.

In the European version and, later, the U.S. version (Fig. 10-11), Philips of the Netherlands has developed a digitally-controlled I^2 C bus so that various units may be turned on and off and otherwise directed. The end function control will allow one or more consumer products to be programmed according to convenient use, especially quasi-split sound receivers whch some U.S. types will use also. Note the two demodulators in the U.S. illustration, followed by multiplex and SAP (second channel) decoders. Philips says that the "first stage of the audio processor is, therefore, a switch for selecting internal or external audio signals . . . the second stage is a switch for language selection." These new techniques allow considerable reduction of peripheral components for pseudo stereo circuits and controls such as volume, balance and tone have been simplified. "For the first time," declares Philips, "it is now economically feasible to process hi-fi sound signals in TV sets."

Zenith's Stereo TV Decoder

Once again, we'll have to do a "wing and a

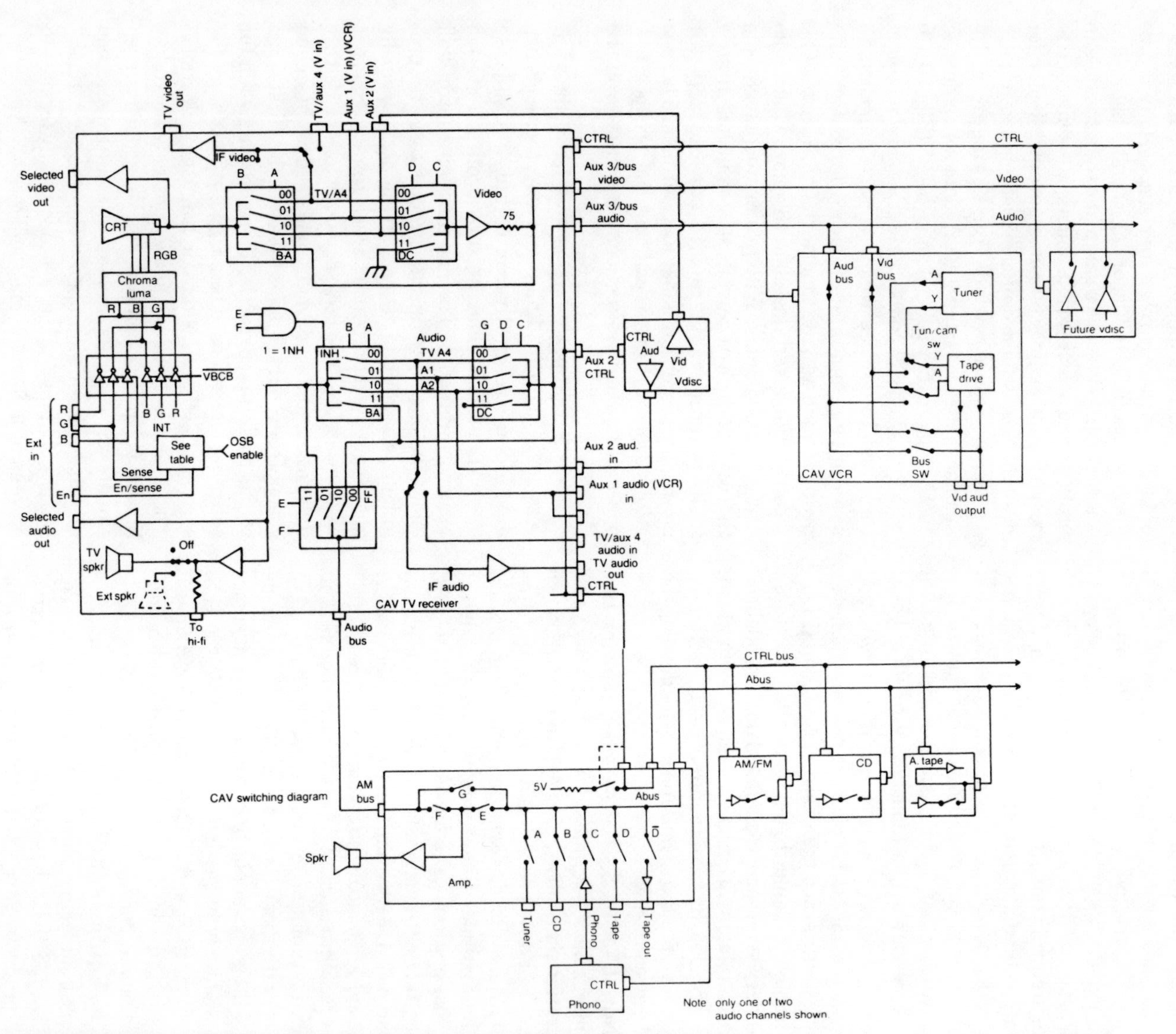

Fig. 10-10. RCA's multiple connector bus that accommodates over a dozen separate equipments—all remotely controlled (courtesy RCA Consumer Electronics).

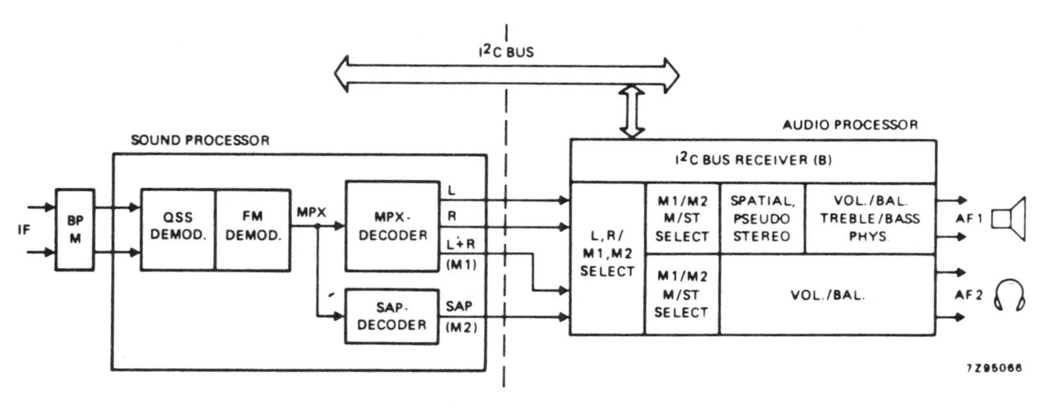

Fig. 10-11. Philips sound processing USA multiplex I²C bus control (courtesy Philips).

prayer" on this Zenith special CV524 stereo adapter since training and artwork aren't fully developed. Nonetheless, there will be enough details in the description to at least give you an idea of its functions. As you will see, this stereo adapter is a rather large and complex piece of equipment (Fig. 10-12) which is not easily characterized in a few sentences, and its 2¼ H × 16 13/16 W × 13⅜ D dimensions aren't precisely minute. You also have to take into consideration that it has both 41.25 MHz carrier *and* 4.5 MHz intercarrier sound inputs, and jacks for two fixed L and R audio outputs at 435 mV, and variable level left and right audio outputs for two 8-ohm external speakers at 2 × 2W. There are also bass and treble tone controls, and max. signal-to-noise ratios of 65 dB—which is pretty good in any set of specifications. In the stereo mode, channel separation amounts to more than 25 dB, distortion less than 1 percent, and 3 dB bandwidth from 80 Hz to 12 kHz. The SAP channel goes from 100 Hz to 8 kHz, with S/N of 50 dB. Mode indicators are light emitting diodes (LEDs).

On the front panel you can see volume up/down, auxiliary stereo, extended stereo, stereo, and monophonic. Stereo and second language (SAP) information picked up by the television receiver is reduced by the TV tuner's first detector to 45.75 MHz video and 41.25 MHz audio carriers which are separated just after the tuner and decoded either as carrier or 4.5 MHz intercarrier audio, depending on the tapoff point. Zenith in its earlier video/audio split "sound" had the former, and later models will feature the quasi-split sound which the rest of the industry has accepted. At best we know, here are the Zenith multichannel receiver identifications:

Y Line (1984) Top End has 41.25 MHz available with external jacks.

Z1, Z2 Chassis All will accommodate 4.5 MHz adapter kits.

Z2 Chassis Top End also has 41.25 MHz jacks.

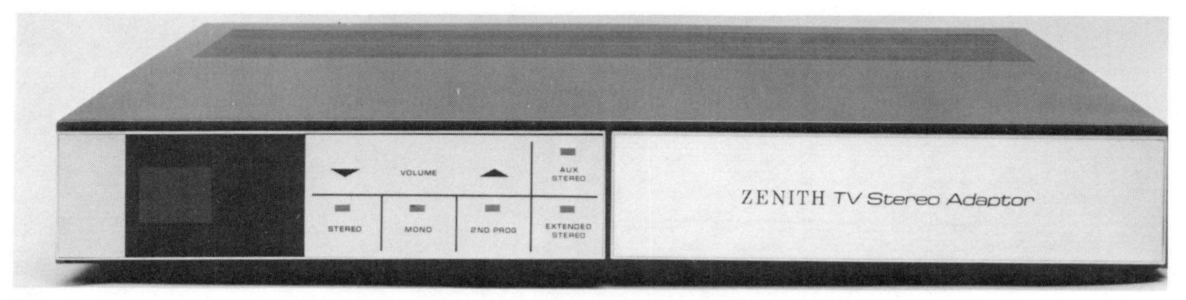

Fig. 10-12. Front view of Zenith's CV524 stereo/SAP adapter (courtesy Zenith Electronics).

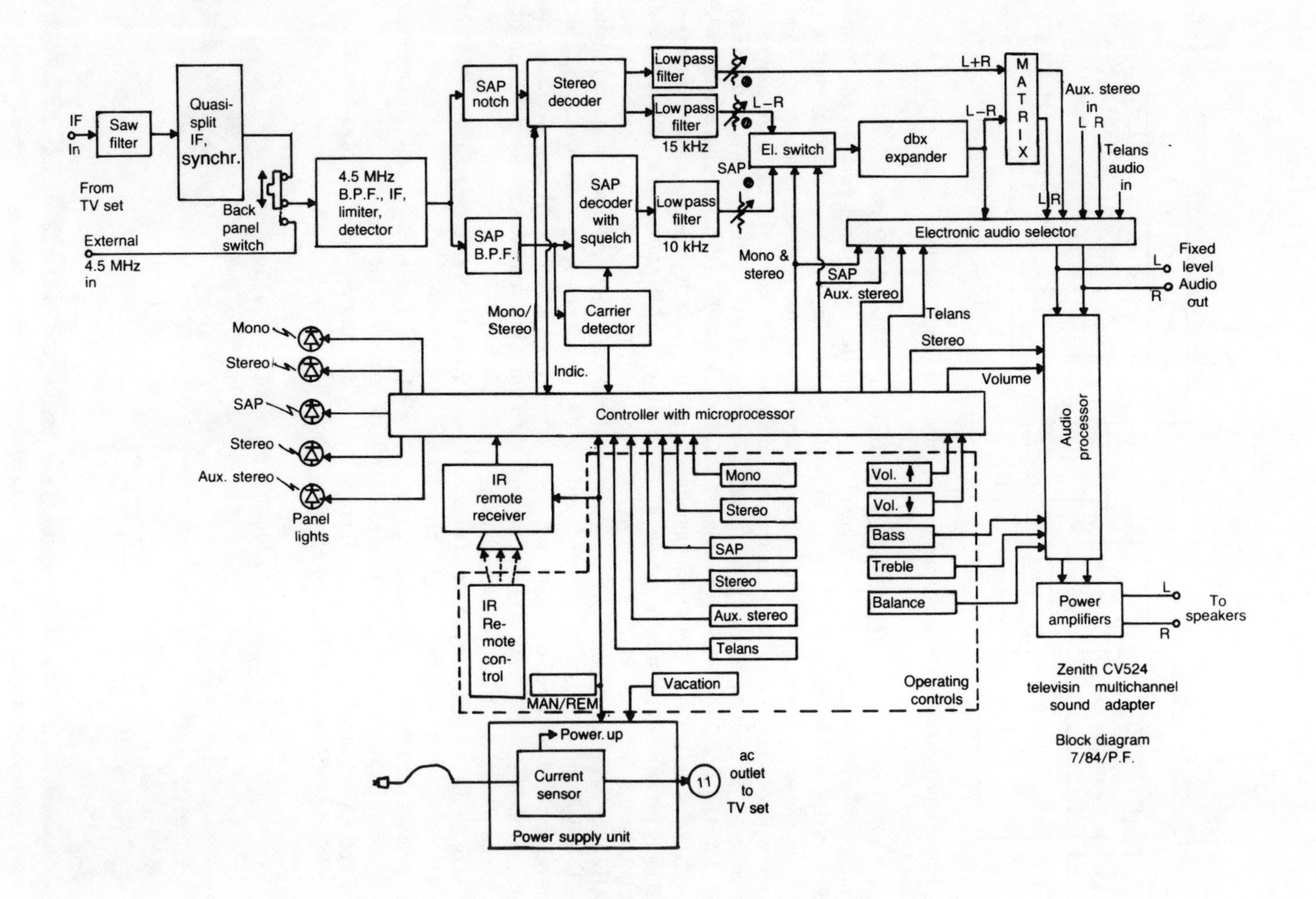

Fig. 10-13. Zenith's stereo decoder/SAP interconnect diagram (courtesy Zenith Electronics).

A Line (1985) Top End has 41.25 MHz jacks and accepts 4.5 MHz adapter kits.

Theory of Operation. The CV524 is microprocessor controlled and operates from either the 41.25 MHz audio carrier or 4.5 MHz intercarrier (Fig. 10-13). Audio information, volume levels or mute may be selected either by membrane keyboard or through an infrared remote transmitter. External stereo inputs can be chosen as well as stereo/SAP, and an automatic "default" changes from SAP to TV stereo if SAP is not on the air. With no stereo, monaural is automatically selected. The ON function takes place when a television receiver connected to the CV524's ac receptacle draws current. Otherwise the unit remains off.

LED lamps light when the various keys are depressed or when the unit automatically chooses SAP, mono, or stereo system operation. There is also a telephone input which has priority over all other operations. Should there be an incoming or outgoing call, the original indicator lights will remain glowing while the conversation is in progress. On phone-equipped receivers, however, an external connection between the set and CV524 is required. Should the *extended stereo* function be desired, a separate LED indicator turns on and stereo speakers will sound as though there's greater distance between them.

Sony's Multichannel System

Just as in AM stereo portables, Sony Corp is also offering a line of television receivers and Profeel monitors that are "stereo ready" and can be readily attached through phone jacks to their multiplex (MPX) multichannel stereo/SAP adapters. Available are receivers in 15, 17, 19, 21, and 26-inch screen sizes, and one 19 and one 25-inch Profeel monitor(s). Beta hi-fi VCRs so equipped are also included.

The MLV-1100 adapter, requiring 35 watts ac power and measuring 17 W × 2⅝ H × 11⅛ D, has one MPX phone jack input, as well as two LINE (another audio source) inputs; two equivalent LINE outputs at −5 dB (435 mV rms) delivering load impedances of 4.7 kilohms. There's also a headphone jack for both low and high impedance stereo headphones (Fig. 10-14), and an unswitched convenience ac outlet for other audio/video equipments. You will also find a built-in amplifier permitting direct connections to hi-fi or Profeel component speakers of 4-8 ohms.

Unlike some of its contemporaries, Sony multichannel sound is not tapped near the tuner but after limiting and FM detection (Fig. 10-15), pass through an emitter-follower current and coupling driver transformer to the MPX jack output on an antenna terminal board. A large capacitor and primary-secondary windings of the untuned transformer securely isolate the output from any hot chassis hazard or undesirable dc component source.

When multichannel audio is received, auto stereo automatically flags MLV-1100 and operates the decoder. On the front panel are Main, SAP, or both buttons permitting selection of stereo, SAP, or a combination of the two when you are interested in the two signals combined in monaural, with indicator lamps identifying your selection.

There are balance, treble, bass, and volume controls available, too, in addition to a LINE-in button and lamp. Input-output jacks and LEVEL signal amplitude adjust are on the back. At the moment, there are no functional diagrams of the multichannel decoder available and probably won't be for some months to come. Considering other BTSC writeups preceding, it will have to do exactly what everyone else has done, with the composite baseband exceptions since the FCC has authorized only one BTSC-dbx system with the main carrier at 15,734 Hz.

General Electric also will market a composite baseband stereo system into the BTSC-dbx decoder using a TDA2546 quasi-split sound and detector chip to do the job. Whether these "different" systems will induce more buzz and less stereo separation remains to be seen, and we really won't know until broadcasters are delivering signals over the air. But you should pay attention to the two systems since they are on the market for consideration and comparison. Certainly Sony's MPX stereo ready jack will easily handle audio signals through 100 kHz within about 8 dB when measured with a

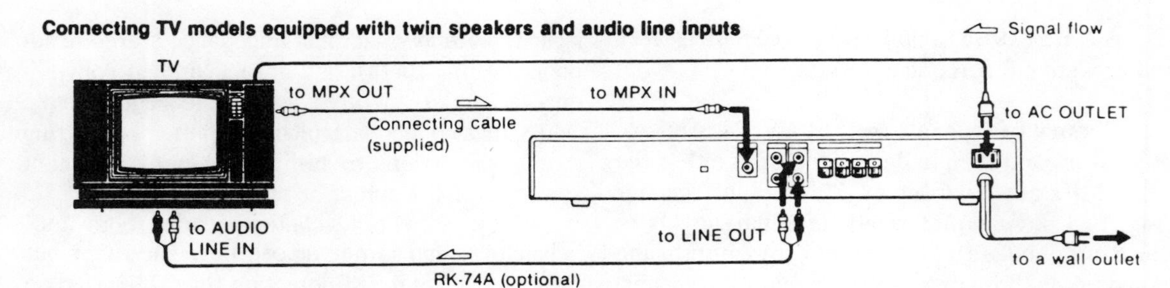

Regulate the sound level with the TV's volume control.

Connecting a component TV and a TV tuner

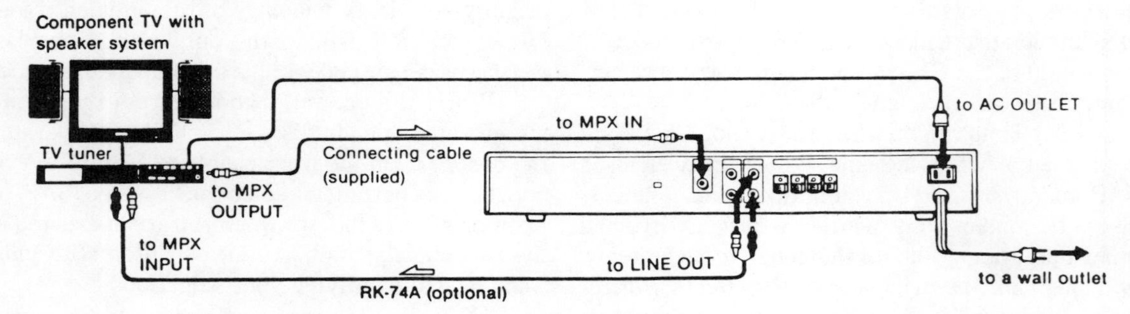

Regulate the sound level with the TV tuner's volume control.

Connecting TV models equipped with one speaker
To enjoy full stereo separation, you will need a pair of stereo speakers.

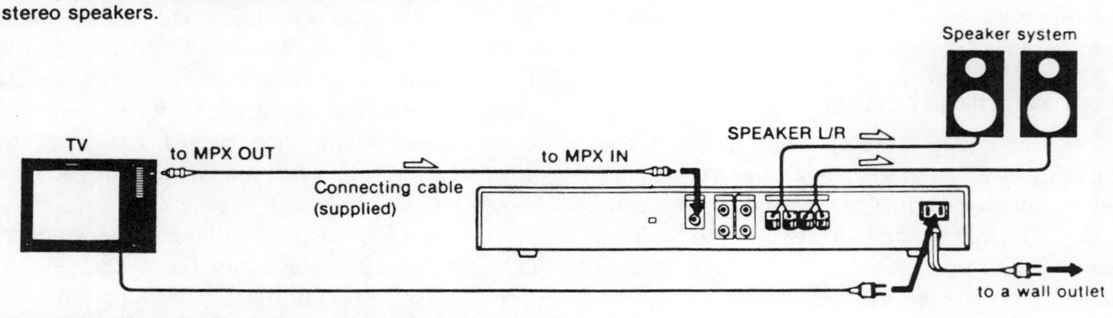

Turn off the TV's volume control and regulate the sound level with the VOLUME control of the MLV-1100.

Fig. 10-14. An example of Sony's stereo adapter connections (courtesy Sony Corp.).

spectrum analyzer at the MPX output.

The Magnavox Stereo TV System

At least in the first go-round, Magnavox (North American Philips) will be going with TDA European ICs for basic stereo sound, taking advantage of several years of prior experience with the medium. The method, however, turns out to be the usual

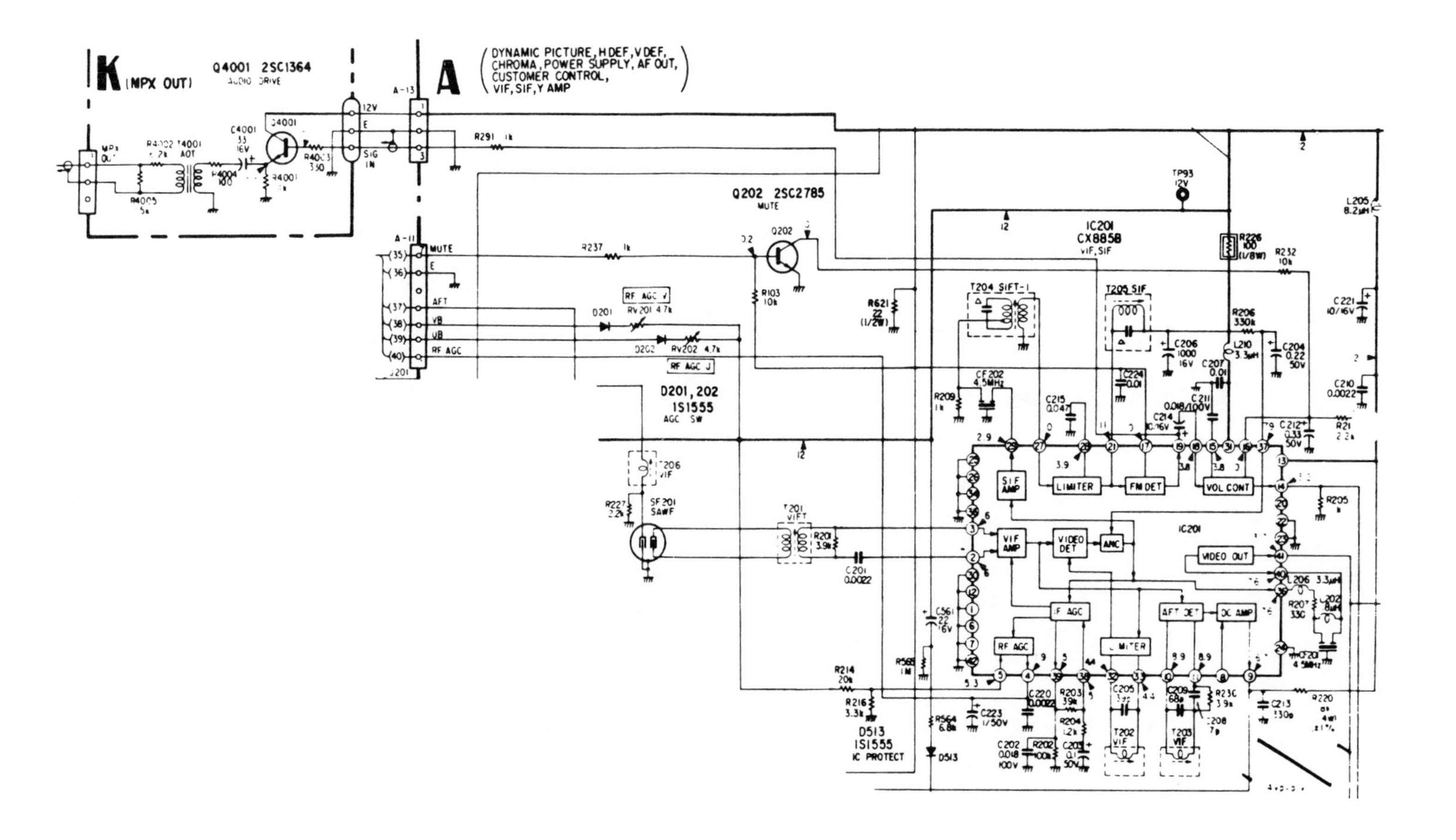

Fig. 10-15. Schematic of video/audio IF and stereo sound takeoff at the FM detector (courtesy Sony Corp.).

quasi-split version and is the same approach we're using in the States but adding, of course, the dbx noise reduction and expander. Advanced development television chassis engineering manager Lee Hoke, Jr. describes the 4.5 MHz sound intercarrier processes as the best we've seen and worthy of your attention. Part of the video and sound carriers, he says, are routed through a "double-peaked" bandpass circuit with a trap at the 42.17 MHz chroma subcarrier frequency. Picture and sound carriers are now on flat response curves so that virtually no ICPM (Nyquist incidental carrier phase modulation) is present.

Audio is then detected usually at quadrature rather than by simple envelope, producing a much more linear and satisfactory output. A further European comment claims that since there is no attenuation of the sound carrier, S/N ratios of the system are improved over conventional ICs where audio is detected just *after* the IF amplifiers. Also, since bandpass amplifiers specifically pass the separate video and sound information immediately after the SAW filters, wider bandpasses for each can be expected with less mixing problems.

Probaby the best way to illustrate what Magnavox does, is to copy portions of its new 25C5 chassis schematic and run through the called-out IC functions in light detail.

Taking its dual input from the tuner, IF preamplifier, and surface wave acoustical filter (SAW), midpoint 44 MHz video, audio, sync, and chroma pass into a gain-controlled amplifier, the outputs of which proceed to a sync detector and tuned reference (REF) amplifier. This stage then produces intelligence around the video carrier for the automatic fine tuning detector, as well as a sync-demodulator input (Fig. 10-16).

A video preamplifier then increases the output of these operations, supplying signals for the white spot inverter as well as the combined noise inverter and AGC detector, whose output is returned to gain-control the IF amplifier and, through an rf delay potentiometer, to the tuner rf amplifier. Automatic fine tuning also has its tank circuit and swings about 45.75 MHz video, furnishing a well-filtered dc control voltage for the tuner oscillator, keeping it on channel track. Audio takeoff, as you see, originates with the video preamplifier and is peaked and buffered by an emitter follower transistor for the 4.5 MHz output, the intercarrier beat heterodyne seemingly occurring somewhere in the video preamplifier, although this function is not specifically identified. The white spot inverter could be an overmodulation check to prevent undue white amplitudes in certain scenes, which would adversely affect luminance in the picture.

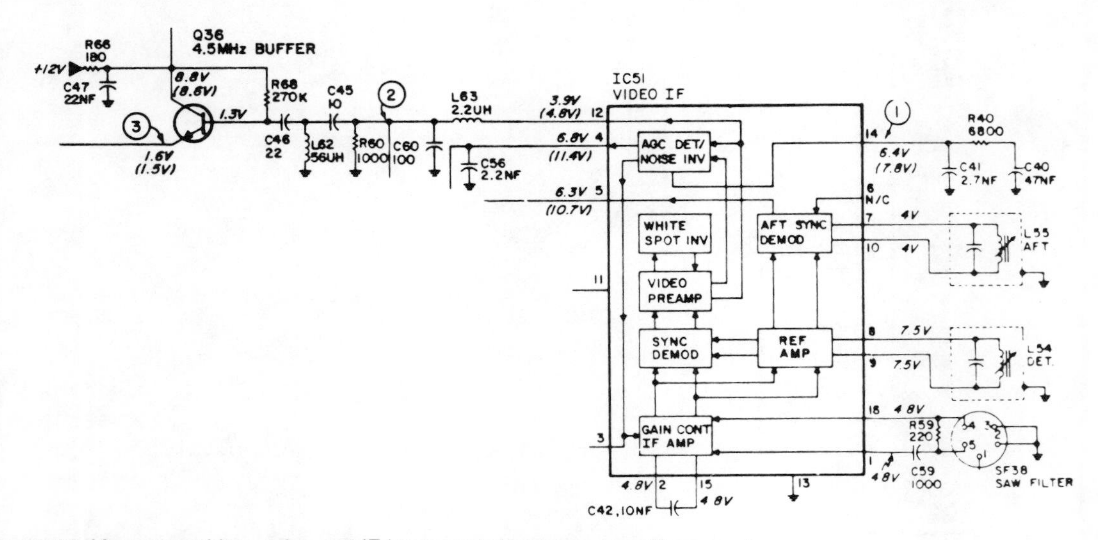

Fig. 10-16. Magnavox video and sound IF integrated circuit (courtesy Magnavox).

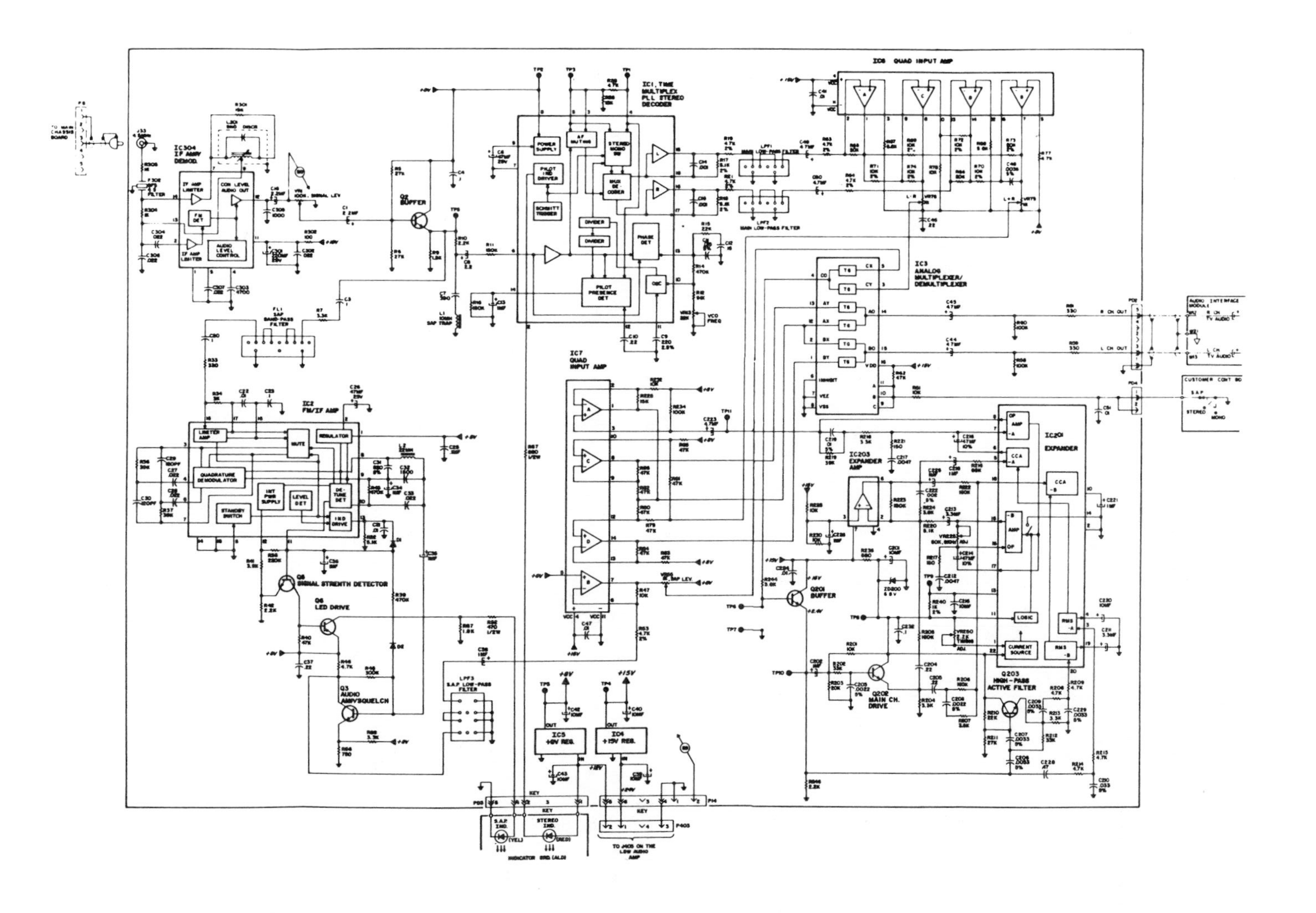

Fig. 10-17. Main chassis portion of the Magnavox stereo sound decoder (courtesy Magnavox).

We now move from the main chassis to the stereo decoder panel and the rest of the stereo/SAP processing. Here, in IC304, you see the IF audio limiter amplifiers, the FM detector and the RLC sound discriminator. A dc signal level is then potentiometer-set, and a low impedance buffer-driver couples signals through to either the multiplex phase-locked loop stereo decoder or through the SAP bandpass filter to the second audio program (SAP) circuits below (Fig. 10-17).

IC1 consists of 14 block functions, three of which are amplifiers for inputs and outputs. As for the rest, there are the familiar vco oscillator, phase detector, pilot detector, two sets of dividers, a Schmitt trigger for mono/stereo switching, pilot lamp indicator driver, power supply, audio muting, and the multiplex decoder. You also have a vco rheostat dc frequency adjust, an RC phase network for the phase detector, plus oscillator and power supply filters. The two outputs are stereo right and left or plain monophonic signals.

SAP Channel. If the carrier frequency is $5f_h$ or 78.67 kHz, the separate audio program channel bandpass filter allows this FM information to pass and frequency demodulation detection takes place in IC2. Incoming lower frequency signals up to 10 kHz are limited, muted (if required) and passed on to the quadrature demodulator, which then supplies the regulated detuning detector, in addition to squelch and other outboard signal strength and LED driver transistors. A level adjust sets the independent drive parameters, and a standby switch probably disables it under certain non signal conditions. IC6 and IC7 quad amplifiers now receive stereo/mono or SAP from the two ICs and then each is processed by dbx expander—types IC201/IC203—reminiscent (or possibly indentical) to the rms level sensor and voltage-controlled amplifier *NEC dbx* described earlier in the chapter. Not having a precise pinout of the chips we aren't certain, but the principle should be identical, as the dynamic range compressor in the transmitter encoder becomes a range expander in the decoder by a factor of two. Note the active high pass filter and its RC time constants, in addition to the buffer and main channel driver.

Final outputs are subsequently processed as full right channel and left channel signals and SAP to the final amplifiers elsewhere. Just below these outputs observe the customer-controlled SAP, stereo, mono switch, and its three selective positions.

CONCLUSION

With multichannel sound systems described for RCA, Zenith, Sony, and North American Philips (Magnavox, Sylvania, and Philco), this entire subject should be reasonably well wrapped up and fairly presented. Fortunately or unfortunately, there's never an end to contemporary information, and new designs and approaches are developed every day. Sorry we can't include them all, but you'll encounter the principles for a long time to come, as well as the actual circuits in this book. Without the enthusiastic and considerate cooperation of training, publications, engineering, and excellent public relations, this would never have happened.

Now it's up to everyone to see that a new industry enjoys a running start and thrives, despite some poor design work here and there, and forthcoming field experiences that could not have occurred without actual equipment operating in consumer homes. In the end, and that means only a couple of years wait, stereo TV sound will become a universal demand item, along with full video/chroma response receivers resulting in delightful viewing and listening for everyone. We would hope, also, that AM stereo for radios enjoys at least some of this great consumer popularity. Both of these developments are good for us all, our highly competitive industry, and the financial well-being of America. Recessions and depressions don't flourish in times of progress. And progress doesn't occur unless there's a need, and something to fill that need. We sincerely hope these needs are AM radio stereo and television multichannel sound!

Appendix A

FCC Evaluation Criteria

THE SOURCES OF THE ENGINEERING DATA USED AND THE SCORING CRITERIA USED IN THE COMPLETION OF THE AM STEREO SYSTEM EVALUATION TABLE

A. SOURCES OF THE ENGINEERING DATA

I-1 MONOPHONIC COMPATIBILITY (HARMONIC DISTORTION)

MAGNAVOX Magnavox Comments of February 9, 1981, page III-4, Station WOWO.

BELAR Belar Comments of May 29, 1981, pages 38, 39 and 43. These were laboratory data. Belar did provide us with over-the-air measurements of Station WJR. However, they did not feel that these measurements adequately reflected the performance of their system in this category.

II-1 INTERFERENCE CHARACTERISTICS (OCCUPIED BANDWIDTH)

MAGNAVOX NAMSRC Report of December 1977, pages H-93, H-96, H-99 and H-102.

MOTOROLA NAMSRC Report of December 1977, pages H-94, H-97, H-100 and H-103.

HARRIS Harris Comments of May 15, 1979, Appendix VI, Figures 1-4, 2-4, 3-4 and 5-4.

BELAR — Belar Comments of May 29, 1979, page 9.

KAHN — Kahn Comments of May 14, 1979, Figure 17-c (42.5%, R-only, 85%, L=R, 85%, L=R), Figure 17-f (42.5%, L-only).

II-2 — INTERFERENCE CHARACTERISTICS (PROTECTION RATIOS)

BELAR — Belar did not submit comments on the Further Notice. As we stated in paragraph 28 of the Further Notice, we could only score the Belar system as 1 on the data which we had at that point.

IV-4 — TRANSMITTER STEREO PERFORMANCE (NOISE)

KAHN — Kahn Comments of February 9, 1981, page 16.

V — RECEIVER STEREO PERFORMANCE

MAGNAVOX — Magnavox Comments of February 9, 1981, Figure 20, Station WOWO, Philips 673#6 Receiver, 42.5% L-only, R-only modulation.

MOTOROLA — Motorola Comments of February 9, 1981, Part 2, page 124, Figures 18 and 19, Station WTAQ, Philips AH6731 receiver, 42.5% L-only, R-only.

HARRIS — Harris Comments of February 9, 1981, Appendix I, pages 8 and 11, Station WGEM, Sansui receiver, 50% L-only, R-only.

BELAR — Belar did not submit Comments on the Further Notice. We gave the Belar system an average score of 5 in this category.

KAHN — Neither Kahn nor Hazeltine furnished us with the data with which we could score this category. We gave the Kahn/Hazeltine system an average score of 5 in this category.

B. SCORING CRITERIA USED IN COMPLETING THE EVALUATION TABLE

In completing the AM Stereo System Evaluation Table, we used the same scoring method that was used in the Further Notice wherever this was possible. For some of the categories no scoring criteria had been established. In the following paragraphs, we explain our method of scoring these categories.

II-1 — INTERFERENCE CHARACTERISTICS (OCCUPIED BANDWIDTH)

The highest modulating frequency used in the 4-tone test is 9500 Hz. If no harmonics of these 4 tones or their intermodulation products are generated by the system, then all of the spectral components will lie within a 20 kHz band centered about the carrier. The presence of spectral components outside this region indicates that, in the case of stereo transmissions, a larger bandwidth is required than that needed to accommodate the basic modulation tones alone. The magnitude of these emissions provides an indirect measurement of the bandwidth occupancy of the system.

Our method of scoring this category is relatively simple and is similar to the one used by NTIA in their evaluation of occupied bandwidth in their Comments of January 10/February 16, 1978. For each of the four different stereo transmission conditions (L=R, L-R, L-only, R-only), we looked for the largest single spectral component that was generated outside the 20 kHz band. In each of the four cases the magnitude of the component was determined relative to the carrier. The average of the four values was determined and a score was assigned on the basis of the following table:

Spectral Component Level (-dB)	Points
67-70	10
64-67	9
61-64	8
58-61	7
55-58	6
52-55	5
49-52	4
46-49	3
43-46	2
40-43	1
less than 40	0

For one of the proponents, all of the measured harmonic and intermodulation products were less than the -70dB measurement limit of the spectrum analyzer which they used. We chose this value as one end of our scoring range. The other limit, -40dB, ws chosen so as to bracket all of the measured values and yet not introduce too great an error by providing too large a scoring increment.

III-1 COVERAGE (STEREO TO MONO RECEIVER)

We awarded all five systems under consideration the maximum number of points, 5, in this category. All of the five systems are designed to amplitude modulate the transmitted stereo signal up to the maximum modulation limits of +125% and -100%. The coverage data supplied by the proponents confirms this. However, in order to achieve these levels of modulation, receiver blanking will be required for four of these systems on the -100% modulation peaks.

V RECEIVER STEREO PERFORMANCE

In the Further Notice it was proposed that this category be scored partly on the basis of the receiver's average stereophonic frequency response, separation, distortion and noise characteristics; and partly on the basis of how the system performed under adverse conditions such as deep fading, reception in or adjacent to null areas of directional arrays, and co-channel and adjacent channel interference. However, much of the data which the proponents supplied in regard to these categories were of a qualitative nature and could not be used for scoring purposes. Only the data supplied on stereo distortion and separation could be used as a means of rating the stereo receiver performance.

The data supplied on frequency response were not used as evaluative critera since it was believed that the frequency response of a system can always be adjusted or corrected by appropriate design improvements or equalizers in the transmitter or receiver.

Sony has submitted measurements on the receiver performance of the Magnavox, Motorola, Harris, Belar and Kahn systems as a function of receiver tuning. However, they used only a few selected frequencies (400 and 1000 Hz) and modulations (45% mono, 30% stereo). We do not believe that these frequencies and modulations are representative of actual broadcast conditions. Because of this, we were unable to make use of their data in our scoring of this category.

The measurements of distortion and separation which were supplied by the proponents at frequencies of 400, 1000, 2000, 3000 and 5000 Hz were averaged. The averages were converted to scores based on the following tables.

Distortion (%)	Points
0-1	5
1-2	4
2-3	3
3-4	2
4-5	1
greater than 5	0

Separation (-dB)	Points
25 or better	5
23-25	4
21-23	3
19-21	2
16-19	1
less than 16	0

Our method of scoring these two categories was consistent with the ways Category I-1, Average Transmitter Harmonic Distortion and Category IV-3, Transmitter Stereo Separation had been scored.

Appendix B

FCC Standards for AM Stereo Broadcast

1. Section 2.983 is amended by adding new paragraph (j) to read as follows:

§2.983 Application for type acceptance.

*   *   *   *   *

(j) An application for type acceptance of an AM broadcast stereophonic exciter-generator intended for interfacing with existing type-accepted transmitters shall include measurements made on a complete stereophonic transmitter. The instruction book required under paragraph (d)(8) of this Section shall include complete specifications and circuit requirements for interconnecting with existing transmitters. The instruction book must also provide a full description of the equipment and measurement procedures for performing equipment performance measurements and modulation monitoring to determine that the combination of stereo exciter-generator and transmitter meet the minimum specifications given in §73.40

2. Section 2.989 is amended by redesignating paragraphs (e)(2), (3), (4), and (5) as (e)(3), (4), (5), and (6); revising paragraph (e)(1), and adding new paragraph (e)(2) to read as follows:

§2.989 Measurement required: Occupied bandwidth.

*   *   *   *   *

(e) * * *

(1) AM broadcast transmitters for monaural operation--when amplitude modulated 85% by a 7,500 Hz input signal.

(2) AM broadcast stereophonic operation--when the transmitter operated under any sterephonic modulation condition not exceeding 100% on negative peaks and tested under the conditions specified in §73.128 in Part 73 of the FCC rules for AM broadcast stations.

* * * * *

3. Section 2.1001 is amended by adding new paragraph (d) to read as follows:

§2.1001 Changes in type accepted equipment.

* * * * *

(d) The interfacing of a type accepted AM broadcast stereophonic exciter-generator with a type accepted AM broadcast transmitter in accordance with the manufacturer's instructions and upon completion of equipment performance measurements showing that the modified transmitter meets the minimum performance requirements applicable thereto is defined as a Class I permissive change for compliance with this Section.

4. Section 73.14 is amended by revising the definitions of "AM broadcast channel;" deleting the definitions of "Modulated stage" and "Modulator stage;" and adding new definitions for "Amplitude modulated stage," "Amplitude modulator stage,""Incidental phase modulation," "Left (or right) signal," "Left (or Right) stereophonic channel," "Main channel," "Stereophonic channel," "Stereophonic crosstalk," "Stereophonic pilot tone," and "Stereophonic separation" to read as follows:

73.14 AM broadcast definitions.

* * * * * * * * *

AM broadcast channel. The band of frequencies occupied by the carrier and the upper and lower sidebands of an AM broadcast signal with the carrier frequency at the center. Channels are designated by their assigned carrier frequencies. The 107 carrier frequencies assigned to AM broadcast stations begin at 540 kHz and are in successive steps of 10 kHz to 1600 kHz.

* * * * * * *

Amplitude modulated stage. The radio-frequency stage to which the modulator is coupled and in which the carrier wave is modulated in accordance with the system of amplitude modulation and the characteristics of the modulating wave.

* * * * * *

Amplitude modulator stage. The last amplifier stage of the modulating wave which amplitude modulates a radio-frequency stage.

* * * * *

Incidental phase modulation. The peak phase deviation (in radians) resulting from the process of amplitude modulation.

*   *   *   *   *

Left (or right) signal. The electrical output of a microphone or combination of microphones placed so as to convey the intensity, time, and location of sounds originated predominately to the listener's left (or right) of the center of the performing area.

*   *   *   *   *

Left (or right) stereophonic channel. The left (or right) signal as electrically reproduced in reception of AM stereophonic broadcasts.

*   *   *   *   *

Main channel. The band of audio frequencies from 50 to 15,000 Hz which amplitude modulates the carrier.

*   *   *   *   *

Stereophonic channel. The band of audio frequencies from 200 to 15,000 Hz containing the stereophonic information which modulates the radio frequency carrier.

*   *   *   *   *

Stereophonic crosstalk. An undesired signal occuring in the main channel from modulation of the stereophonic channel or that occuring in the stereophonic channel from modulation of the main channel.

*   *   *   *   *

Stereophonic pilot tone. An audio tone of fixed or variable frequency modulating the carrier during the transmission of stereophonic programs.

*   *   *   *   *

Stereophonic separation. The ratio of the electrical signal caused in the right (or left) stereophonic channel to the electrical signal caused in the left (or right) stereophonic channel by the transmission of only a right (or left) signal.

*   *   *   *   *

5. Section 73.40 is amended by revising the introduction of paragraph (a) and adding new paragraph (b) to read as follows:

§73.40 AM transmission system performance requirements.

(a) The design, installation, and operation of a monophonic AM broadcast transmission system between a common audio input amplifier at the studio to the transmitting antenna terminals must meet the following specifications.

*   *   *   *   *

(b) The design, installation, and operation of a stereophonic AM broadcast transmission system between audio input amplifers used for all programming for both the left and right program channels to the transmitting antenna terminals must meet the following specifications.

(1) Except when due to equipment failures or other conditions beyond the licensee's control, the transmitter must be capable of operating at the authorized power for each mode of operation, with a main (L+R) channel amplitude modulation not less than 85% and a left (L) only or right (R) only signal of not less than 75% over the audio frequency range from 50 to 5,000 Hz.

(2) For main channel modulation only, the transmission system shall meet the distortion specifications of paragraph (a)(2) above with harmonics observed to 20,000 Hz. When stereophonic transmission is used, the distortion must be measured in the left and right channels separately modulated using a suitable stereophonic demodulator.

(3) The audio frequency transmitting characteristics for main (L+R), left (L) only and right (R) only modulation shall conform to the requirements of paragraph (a)(3) above, except that measurements shall extend to 7,500 Hz.

(4) The carrier-amplitude regulation (carrier shift) at any percentage of amplitude modulation by a main (L+R) channel signal shall not exceed 5%.

(6) The carrier hum and extraneous noise level, unweighted noise, over the frequency band 50 to 20,000 Hz for main channel (L+R), left (L), and right (R) channels must be at least 45 dB below the reference level of 100% amplitude modulation of the carrier by a 400 Hz tone. Measurements shall be made with a suitable stereophonic demodulator.

(7) The incidental phase modulation of the transmitter must be measured with the main (L+R) channel modulated at the audio frequencies and modulation levels specified in paragraph (b)(2) above.

Note: Specifications for incidental phase modulation are not established.

(8) For the first five years following installation of stereophonic transmitting equipment, stereophonic separation between left and right stereophonic channels must be at least 15 dB at audio modulating frequencies between 400 and 5,000 Hz. After five years, stereophonic separation between left and right stereophonic channels must be at least 20 dB at audio modulating frequencies between 300 and 5,000 Hz.

6. Section 73.56 is amended by adding new paragraph (e) to read as follows:

§73.56 Modulation monitors.

*   *   *   *   *

(e) AM stations engaged in stereophonic broadcasting shall have sufficient monitoring equipment to insure that the transmitted signal conforms to the modulation requirements of §73.1570 and as specified by the manufacturer of the type accepted stereophonic transmitting equipment in use. The indications of the main channel modulation level during stereophonic transmissions must be available to the transmitter operator as specified in paragraph (a) of this Section.

Note: Requirements for the type approval of AM stereophonic modulation monitors and for the stereophonic modulation indications to be available to the operator are not being established pending Commission action in BC Docket No. 81-698.

7. New Section 73.128 is added to read as follows:

§73.128 AM stereophonic broadcasting.

(a) An AM broadcast station may, without specific authority from the FCC, transmit stereophonic programs upon installation of type accepted stereophonic transmitting equipment and the necessary measuring equipment to determine that the stereophonic transmissions conform to the modulation characteristics specified for the stereophonic transmission system in use.

(b) The FCC does not specify the composition of the transmitted sterephonic signal. However, the following limitations on the transmitted wave must be met to insure compliance with the occupied bandwidth limiations, compatibility with AM receivers using envelope detectors, and any applicable international agrements to which the United States is a party:

(1) The transmitted wave must meet the occupied bandwidth specifications of 73.44 under all possible conditions of program modulation. Compliance with requirement shall be demonstrated either by the following specific modulation tests or other documented test procedures that are to be fully described in the application for type acceptance and the transmitting equipment instruction manual. (See 2.983 paragraphs (d)(8) and (j)).

(i) Main channel (L + R) under all conditions of amplitude modulation for the sterephonic system but not exceeding amplitude modulation on negative peaks of 100%

(ii) Stereophonic subchannel (L - R) modulated with audio tones of the same amplitude at the transmitter input terminals as in (i) above but with the phase of either the L or R channel reversed.

(iii) Left and Right Channel only, under all conditions of modulation for the stereophonic system in use but not exceeding amplitude modulation on negative peaks of 100%.

(2) The total harmonic distortion as measured by an envelope detector having an input radio frequency bandwidth of 30 kHz (3 dB points) many not exceed 5% for the conditions of modulation specified (1) of this paragraph.

(c) Each licensee or permitee of an AM station engaging in stereophonic broadcasting using a system with a pilot tone shall measure the quiescent pilot tone frequency and injection level and calibrate at intervals as often as necessary to insure compliance with the specifications for the system in use. However, in any event, the measurements shall be made at least once each calendar month with not more than 40 days between successive measurements.

8. Section 73.1545 is amended by revising paragraph (a) to read as follows:

§73.1545 Carrier frequency departure tolerances.

(a) AM stations. The departure of the carrier frequency for monophonic transmissions or center frequency for stereophonic transmissions may not exceed ± 20 Hz from the assigned frequency.

9. Section 73.1570 is amended by revising paragraph (b)(1) to read as follows:

§73.1570 Modulation levels: AM, FM, and TV aural.

(b)

(1) AM stations. In no case shall the amplitude modulation of the carrier wave exceed 100% on negative peaks of frequent recurrence, or 125% on positive peaks at any time.

(i) AM stations transmitting stereophonic programs not exceed the AM maximum stereophonic transmission signal modulation specifications of stereophonic system in use.

(ii) AM stations transmitting telemetry signals for remote control or automatic transmission system operation must meet the modulation limitations of 73.67(b) or 73.142(j).

10. Section 73.1590 is amended by redesignating paragraphs (a)(3), (4), and (5) as (a)(4), (5), and (6); adding new paragraph (a) (3); revising the introduction of paragraph (b) (1); redesignating paragraphs (b)(2) as (b)(3); and adding new paragraph (b)(2) to read as follows:

73.1590 Equipment performance masurements.

(a)***

* * * *

(3) Installation of AM stereophonic transmission equipment pursuant to 73.170.

* * * *

(b)***

(1) AM monophonic stations.

(2) AM stereophonic stations.

* * * *

(i) Data and curves showing the overall audio frequency response from 50 to 15,000 Hz for approximately 25%, 50%, 75%, and 100% modulation with equal equal left and right (L+R) main channel signal; 25%, 50% and 75% modulation with both a left (L) channel only and right (R) channel only signals.

(ii) Data and curves showing audio frequency harmonic content for 25%, 50%, 75% and (main channel only) 100% modulation for the audio frequencies 50, 100, 400, 1000, 5000, and when attinable 7,500, 10,000, 12,500 and 15,000 Hz (either arithmetical or RSS (root sum square) values up to the 10th harmonic or 30,000 Hz) for equal left and right (L=R), left (L) only and right (R) only signals. A family of curves must be plotted as specified in paragraph (b)(1)(ii) above.

(iii) Data showing percentage of carrier amplitude regulation as specified in paragraph (b)(1)(iii) above for main channel modulation with equal left and right (L=R) signals.

(iv) The carrier hum and extraneous noise level generated within the equipment, and measured throughtout the audio spectrum, or band, in dB below the reference level of 100% amplitude modulation by a 400 Hz tone for the main, left, and right channels.

(v) Measurements or observations for spurious and harmonic radiations as specified in paragraph (b)(1)(v) above while modulating the transmitter main (L+R) channel, left (L) channel only, right (R) channel only, and stereophonic channel only (L-R) signal. The tests shall be made with the tones and maximum attainable normal modulation as specified in 73.128(b).

(vi) The degree of incidental phase modulation of the carrier wave, in radians, when the main (L+R) channel is amplitude modulated without pilot tone. The tests shall be made with the tones and maximum attainable modulation levels as specfied in (2)(i) of this paragraph.

(vii) The main to stereophonic channel and the stereophonic to main channel crosstalk. The tests shall be made with the tones and maximum attainable normal modulation as specified in (2)(i) of this paragraph.

(viii) In the above measurements, if 100% negative peak amplitude modulation is not attainable, the highest attainable modulation percentage between 95% and 100% modulation shall be used.

* * * *

(d) AM stereophonic exciter-generators for interfacing with type accepted AM transmitters may be type accepted upon request from any manufacturer by the procedures described in Part 2 of the FCC rules. AM station licensees will not be authorized to use composite or non-type accepted AM stereophonic transmitting equipment under the provisions of paragraphs (b) and((c) of this Section.

(e) ***

12. Section 73.1690 is amended by redesignating paragraph (3)(4) as (e)(5) and adding new paragraph (3)(4) to read as follows:

73.1690 Modification of transmission systems.

* * * *

(e) ***

* * * * *

(4) Modification of the transmitter for stereophonic broadcasting with a stereophonic exciter unit which has been type accepted and designed for interfacing with the type approved transmitter with which it is to be used.

* * * * *

13. The Alphabetical Index - Part 73 is amended by inserting the following new or revised listings in alphabetical sequence:

AM stereophonic broadcasting..............................73.128

Stereophonic broadcasting
- AM...73.128
- FM...73.297
- NCE-FM...73.597

Stereophonic transmission standards
- AM...73.170
- FM...73.322

Appendix C

Listing of AM Stations Currently on the Air

Kahn / Hazeltine Stations

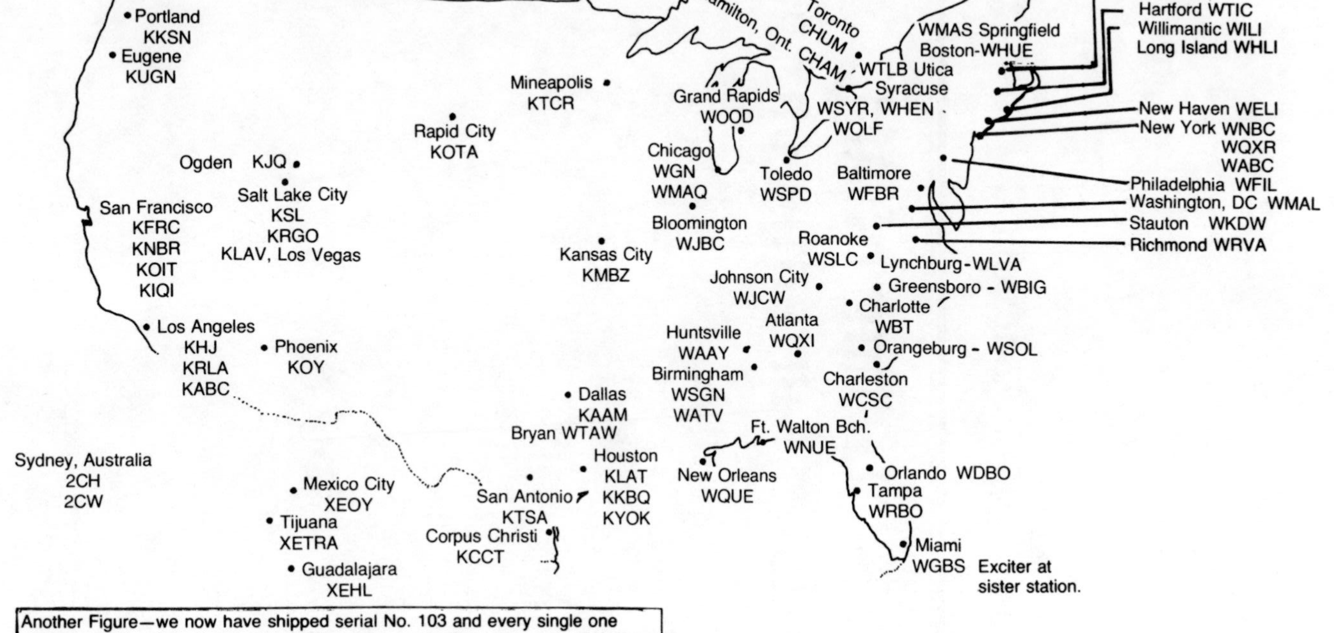
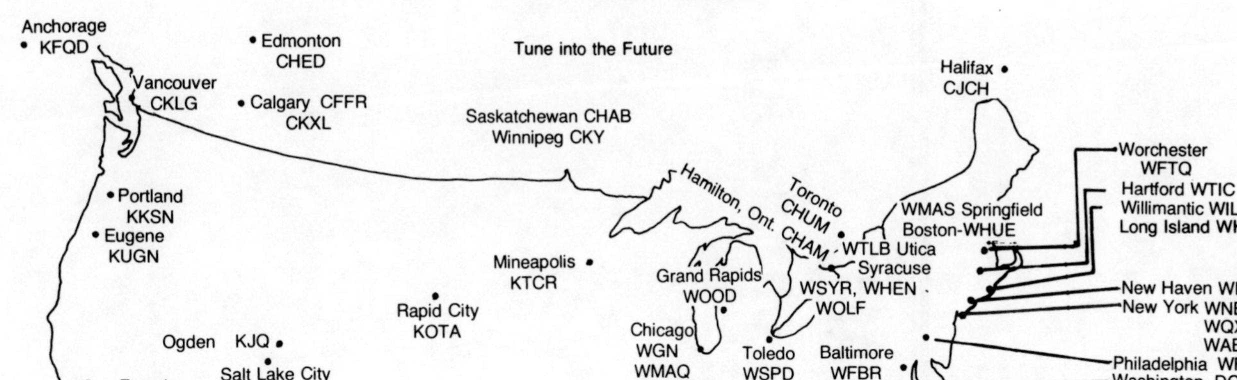

Another Figure—we now have shipped serial No. 103 and every single one of these exciters have been sold, no free loaners or gifts to big or small stations.

Former Harris stations and part time stations not included.

HARRIS EQUIPPED AM STEREO STATIONS

WQXI	Atlanta, Georgia
WESC	Greenville, South Carolina
WNOE	New Orleans, Louisiana
KROW	Reno, Nevada
CKLW	Windsor, Ontario (Detroit)
WGAR	Cleveland, Ohio
WGEM	Quincy, Illinois
WDAE	Tampa, Florida
WCOL	Columbus, Ohio
WLS	Chicago, Illinois
WING	Dayton, Ohio
KFRE	Fresno, California
WBCO	Bucyrus, Ohio
WOOF	Dothan, Alabama
KLRA	Little Rock, Arkansas
WHIZ	Zanesville, Ohio
KIML	Gillette, Wyoming
KOMO	Seattle, Washington
WDAF	Kansas City, Missouri
WSIC	Statesville, North Carolina
WGKA	Atlanta, Georgia
WJMW	Athens, Alabama
KHOW	Denver, Colorado
WPLB	Greenville, Michigan
KYMN	Northfield, Minnesota
KYST	Houston, Texas
WRPQ	Baraboo, Wisconsin
KDAY	Los Angeles, California
WCUZ	Grand Rapids, Michigan
WISE	Asheville, North Carolina
WXOR	Florence, Alabama
KTAM	Bryan, Texas
KJR	Seattle, Washington
WHIN	Gallatin, Tennessee
KXLF	Butte, Montana
KOGA	Ogallala, Nebraska
KSUM	Fairmont, Minnesota
CFRA	Ottawa, Ontario
Neff	Communications, Bethesda, Maryland
CKRB	St. Georges de Beauce, Quebec
KIT	Yakima, Washington
KRSP	Salt Lake City, Utah
KFGO	Fargo, North Dakota
WLAS	Jacksonville, North Carolina
WTUP	Tupelo, Mississippi
KEEN	San Jose, California
KSJL	San Antonia, Texas
KIKR	Conroe, Texas
KAGC	Bryan, Texas
WJXY	Conway, South Carolina
WOW	Omaha, Nebraska
KPRE	Paris, Texas
WBRN	Big Rapids, Michigan
KSO	Des Moines, Iowa
KOGO	San Diego, California
WPAD	Paducah, Kentucky
KRBC	Abilene, Texas
KNOW	Austin, Texas
6IX	Australia
6PM	Australia
WCKB	Dunn, North Carolina
KLOU	Lake Charles, Louisiana
KAIR	Tucson, Arizona
BCNZ	Radio I, New Zealand
Radio Mundo,	Brazil
2KO	Australia
2UE	Australia
2SM	Australia
4BK	Australia
4BH	Australia
4IO	Australia
6PR	Australia
3XY	Australia
2NX	Australia
3MP	Australia
WSOK	Savannah, Georgia
KAGI	Grants Pass, Oregon
KAJO	Grants Pass, Oregon
WQBS	San Juan, Puerto Rico
WORA	Mayaguez, Puerto Rico
WPRP	Ponce, Puerto Rico
WPTF	Raleigh, North Carolina
3AK	Australia
WKIX	Raleigh, North Carolina
WJBO	Baton Rouge, Louisiana
WQHK	Ft. Wayne, Indiana
WGR	Buffalo, New York
CJMS	Montreal, Quebec
CFLS	Quebec City, Quebec
CFRW	Winnipeg, Manitoba
Voz de la Cordillera,	Paraguay
KSJL	San Antonio, Texas
KHEY	El Paso, Texas
WDBQ	Dubuque, Iowa
KCBF	Fairbanks, Alaska
WWWQ	Panama City, Florida
WROZ	Evansville, Indiana
KGFJ	Los Angeles, California
KFYR	Bismarck, North Dakota
KDOK	Tyler, Texas
KATT	Oklahoma City, Oklahoma
KEZY	Anaheim, California

MOTOROLA AM STEREO LISTING

UNITED STATES

STATE	CITY	STATION	FREQ kHz
Alabama	Atmore	WASG	1140
Alabama	Gadsden	WKFX	930
Alabama	Tuscaloosa	WACT	1420
Arkansas	Fayetteville	KFAY	1250
California	Bakersfield	KUZZ	970
California	Los Angeles	KFI	640
California	Los Angeles	KPRZ	1150
California	Los Angeles	KZLA	1540
California	San Fernando	KGIL	1260
California	Palm Springs	KDES	920
California	Sacramento	KRAK	1140
California	San Diego	KFMB	760
California	San Francisco	KSFO	560
California	Stockton	KJOY	1280
Colorado	Denver	KLZ	560
Colorado	Pueblo	KIDN	1350
Colorado	Wray	KRDZ	1000
Florida	Panama City	WDLP	590
Florida	St. Petersburg	WSUN	620
Georgia	Atlanta	WSB	750
Georgia	Atlanta	WPLO	590
Georgia	Columbus	WDAK	540
Hawaii	Honolulu	KIKI	830
Idaho	Boise	KBOI	670
Idaho	Boise	KGEM	1140
Idaho	Idaho Falls	KUPI	980
Illinois	Chicago	WAIT	820
Illinois	Chicago	WGCI	1390
Illinois	East St. Louis	WESL	1490
Illinois	LaGrange	WTAQ	1300
Illinois	Rockford	WROK	1440
Indiana	Indianapolis	WIRE	1430
Indiana	Indianapolis	WNDE	1260
Kentucky	Lexington	WTKC	1300
Kentucky	Louisville	WHAS	840
Kentucky	Pikeville	WPKE	1240
Kentucky	Richmond	WEKY	1340
Louisiana	Alexandria	KSYL	970
Louisiana	Crowley	KSIG	1450
Louisiana	Denham Springs	WLBI	1220
Louisiana	Garyville	WKQT	1010
Louisiana	Lafayette	KXKW	1520
Louisiana	Monroe	KNOE	540
Louisiana	New Iberia	KANE	1240
Louisiana	Rayville	KXLA	990
Louisiana	Shreveport	KOKA	1550
Louisiana	Thibodaux	KTIB	630
Maine	Auburn	WLAM	1470
Maine	Gardiner	WABK	1280
Massachusetts	Boston	WBZ	1030
Michigan	Detroit	WJR	760
Michigan	Lansing	WITL	1010
Michigan	Saginaw	WSAM	1400
Mississippi	Gulfport	WROA	1390
Missouri	St. Louis	KSD	550
Montana	Billings	KGHL	790
Montana	Billings	KOOK	970
Montana	Great Falls	KMON	560
Montana	Kalispell	KOFI	1180
Nebraska	Omaha	KFAB	1110
Nebraska	Omaha	KOIL	1290
Nevada	Las Vegas	KMJJ	1140
New Jersey	Morristown	WMTR	1250
New Jersey	Paterson	WPAT	930
New Jersey	Princeton	WHWH	1350
New Mexico	Albuquerque	KRZY	1450
New York	Buffalo	WKBW	1520
New York	Rochester	WPXY	1280
N. Carolina	Boone	WATA	1450
N. Carolina	Chapel Hill	WCHL	1360
N. Carolina	Charlotte	WSOC	930
N. Dakota	Bismarck	KLXX	1270
N. Dakota	Fargo	KQWB	1550
Ohio	Akron	WAKR	1590
Ohio	Dayton	WONE	980
Oklahoma	Oklahoma City	KXXY	1340
Oklahoma	Tulsa	KRMG	740
Oklahoma	Tulsa	KVOO	1170
Oregon	Eugene	KYKN	1280
Oregon	Hillsboro	KUIK	1360
Oregon	Medford	KYJC	610
Oregon	Pendleton	KTIX	1240
Oregon	Portland	KGW	620
Pennsylvania	Allentown	WSAN	1470
Pennsylvania	Erie	WJET	1400
Pennsylvania	Lancaster	WLPA	1490
Pennsylvania	Reading	WRAW	1340
Pennsylvania	York	WNOW	1250
S. Carolina	Greenwood	WGSW	1350
S. Carolina	Hilton Head Island	WHHQ	1130
S. Dakota	Rapid City	KKLS	920
Tennessee	Kingsport	WKPT	1400
Tennessee	Memphis	WKDJ	980
Tennessee	Nashville	WSM	650
Texas	Corpus Christi	KIKN	1590
Texas	Dallas	KMEZ	1480
Texas	Dallas	KRQX	570
Texas	San Antonio	KKYX	680
Texas	San Antonio	KCOR	1350
Utah	Price	KRPX	1080
Utah	Salt Lake City	KBUG	1320
Utah	Salt Lake City	KFAM	700
Utah	Salt Lake City	KALL	910
Vermont	Burlington	WDOT	1390
Virginia	Alexandria	WPKX	730
Virginia	Danville	WBTM	1330
Virginia	Harrisonburg	WKCY	1300
Virginia	Norfolk	WTAR	790
Virginia	Woodstock	WAMM	1230
Washington	Seattle	KMPS	1300
West Virginia	Charleston	WQBE	950
West Virginia	Huntington	WKEE	800
Wisconsin	Green Bay	WDUZ	1400
Wisconsin	Madison	WISM	1480

CANADA

PROVINCE	CITY	STATION	FREQ kHz
Alberta	Calgary	CHQR	810
Alberta	Calgary	CFAC	960
Alberta	Camrose	CFCW	790
Alberta	Edmonton	CHQT	1110
Alberta	Edmonton	CJCA	930
Alberta	Lethbridge	CHEC	1090
Brit. Col.	Kelowna	CKOV	630
Brit. Col.	New Westminster	CKNW	980
Brit. Col.	Prince George	CJCI	620
Brit. Col.	Vancouver	CJVB	1470
Brit. Col.	Vancouver	CKWX	1130
Brit. Col.	Victoria	CFAX	1070
Brit. Col.	Victoria	CJVI	900
Manitoba	Winnipeg	CKRC	630
Nova Scotia	Sydney	CJCB	1270
Ontario	Hamilton	CHML	900
Ontario	Hamilton	CKOC	1150
Ontario	Kitchener	CKKW	1090
Ontario	London	CFPL	980
Ontario	Ottawa	CJSB	540
Ontario	Toronto	CFRB	1010
Ontario	Windsor	CKLW	800
Quebec	Montreal	CJAD	800
Saskatchewan	Regina	CJME	1300
Saskatchewan	Regina	CKCK	620
Saskatchewan	Saskatoon	CKOM	1250

AUSTRALIA

N.S.W.	Sydney	2WS	1224
Victoria	Melbourne	3AK	1503
Victoria	Melbourne	3KZ	1179
Victoria	Melbourne	3UZ	927
S. Australia	Adelaide	5KA	1197
S. AFRICA	Radio Johannesburg		702
VENEZUELA	Radio Caracas	YVMY	1550

KAHN/HAZELTINE AM STEREO STATIONS

Station	Frequency (kHz)	Location
KFQD	750	Anchorage, AK
WQXI	790	Atlanta, GA
WFBR	1300	Baltimore, MD
WSGN	610	Birmingham, AL
WATV	900	Birmingham, AL
WJBC	1230	Bloomington, IL
WHUE	1150	Boston, MA
WCSC	1390	Charleston, SC
WBT	1110	Charlotte, NC
WGN	720	Chicago, IL
WMAQ	670	Chicago, IL
KCCT	1150	Corpus Christi, TX
KAAM	1310	Dallas, TX
KUGN	590	Eugene, OR
WNUE	1400	Fort Walton Beach, FL
WOOD	1300	Grand Rapids, MI
WBIG	1470	Greensboro, NC
WTIC	1080	Hartford, CT
KLAT	1010	Houston, TX
KYOK	1590	Houston, TX
KKBQ	790	Houston, TX
WAAY	1550	Huntsville, AL
WJCW	910	Johnson City, TN
KMBZ	980	Kansas City, MO
KABC	790	Los Angeles, CA
KHJ	930	Los Angeles, CA
KRLA	1110	Los Angeles, CA
WLVA	590	Lynchburg, VA
WGBS	710	Miami, FL
KTCR	690	Minneapolis, MN
WELI	960	New Haven, CT
WQUE		New Orleans, LA
WABC	770	New York, NY
WNBC	660	New York, NY
WQXR	1560	New York, NY
WOW	590	Omaha, NB
WSOL	1370	Orangeburg, SC
WDBO	580	Orlando, FL
WFIL	550	Philadelphia, PA
KOY	550	Phoenix, AZ
KKSN	910	Portland, OR
KOTA	1380	Rapid City, SD
WRVA	1140	Richmond, VA
WSLC	610	Roanoke, VA
KSL	1160	Salt Lake City, UT
KRGO	1550	Salt Lake City, UT
KFRC	610	San Fransisco, CA
KNBR	680	San Francisco, CA
KTSA	550	San Antonia, TX
WKDW	900	Staunton, VA
WSYR	570	Syracuse, NY
WHEN	620	Syracuse, NY
WOLF	1490	Syracuse, NY
WRBQ	1380	Tampa, FL
WSPD	1370	Toledo, OH
WMAL	630	Washington, DC
WILI	1400	Willimantic, CT
WFTQ	1440	Worcester, MA

FOREIGN STATIONS EQUIPPED WITH KAHN/HAZELTINE SYSTEM

CHED	630	Edmonton, Canada
CHAM		Ontario, Canada
CHUM	1050	Toronto, Canada
CKY	580	Winnepeg, Canada
CJCH	920	Halifax, Nova Scotia
XEHL	960	Guadalejara, Mexico
XEOY	1000	Mexico City, Mexico
XETRA	690	Tijuana, Mexico
2CH		Sydney, Australia
2VW		Sydney, Australia

Appendix D

FCC Definitions and Regulations

TRANSMISSION STANDARDS

Television Stereophonic Sound Standards

Television broadcast stations may transmit stereophonic sound by employing a subcarrier on the aural carrier. The main channel modulating signal shall be the stereophonic sum modulating signal; the subcarrier modulation shall be the stereophonic difference encoded signal.

The subcarrier shall be the second harmonic of a pilot signal which is transmitted at a frequency equal to the horizontal line rate. Note: if the station is engaged in stereophonic sound transmission accompanied by monochrome picture transmission the horizontal scanning frequency shall be 15,734 Hz ± 2 Hz.

The subcarrier shall be double sideband amplitude modulated with suppressed carrier and shall be capable of accepting a stereophonic difference encoded signal over a range of 50-15,000 Hz.

The total modulation of the aural carrier, including that caused by all subcarriers, shall comply with the requirements of S73.1570 of the FCC Rules and Regulations.

Television Second Audio Program Standards

Television broadcast stations may transmit a subcarrier carrying a second audio program.

The subcarrier frequency shall nominally be equal to the fifth harmonic of the horizontal line rate.

The second program encoded signal shall frequency modulate the subcarrier to a peak deviation of ±10 kHz.

The second audio program subchannel shall be capable of accepting second program encoded signals over a range of 50-10,000 Hz.

The modulation of the aural carrier by the second audio program subcarrier shall comply with S(D) (a) (1) (iv) of this bulletin (±15 kHz deviation).

Television Sound Encoding Standards

The stereophonic difference audio signal and the second program audio signal shall be encoded prior to modulating their respective subcarriers. A diagram of one method of obtaining this encoding is shown as Fig. 8-1.

This encoding shall have the following characteristics, where f is expressed in kilohertz (kHz).

(i) Fixed pre-emphasis (F(f)) whose transfer function is as follows:

$$F(f) = \left(\frac{(jf/0.408)+1}{(jf/5.23)+1}\right) \left(\frac{(jf/2.19)+1}{(jf/62.5)}\right) +1$$

(ii) Wideband amplitude compression wherein:

(a) The decibel gain (or loss) applied to the audio signal during encoding is equal to minus one times the decibel ERMS value of the encoded signal (the result of the encoding process), weighted by a transfer function (P(f)) as follows:

$$P(f) = \frac{(jf/0.0354)}{((jf/0.0354)+1) \quad ((jf/2.09)+1)}$$

(b) The exponential time weighting period T_1 of the ERMS detector referred to above in (a) is 34.7 ms.

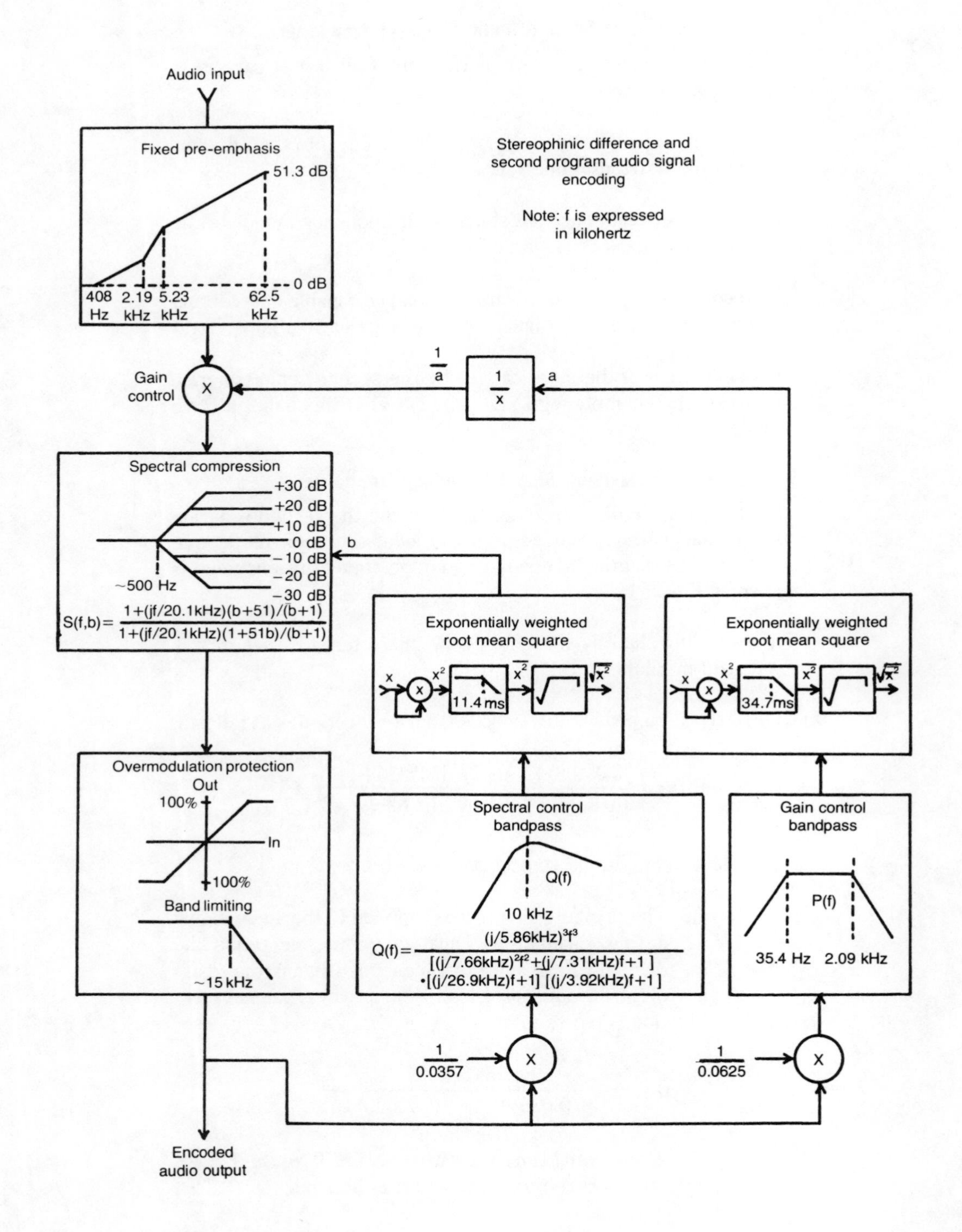
Audio input
Fixed pre-emphasis
51.3 dB
0 dB
408 Hz
2.19 kHz
5.23 kHz
62.5 kHz
Stereophinic difference and second program audio signal encoding
Note: f is expressed in kilohertz
Gain control
X
1/a
1/x
a
Spectral compression
+30 dB
+20 dB
+10 dB
0 dB
−10 dB
−20 dB
−30 dB
~500 Hz
b
S(f,b) = 1+(jf/20.1kHz)(b+51)/(b+1) / 1+(jf/20.1kHz)(1+51b)/(b+1)
Exponentially weighted root mean square
x
x²
11.4 ms
34.7ms
Overmodulation protection
Out
100%
In
100%
Band limiting
~15 kHz
Spectral control bandpass
Q(f)
10 kHz
Q(f) = (j/5.86kHz)³f³ / [(j/7.66kHz)²f²+(j/7.31kHz)f+1] •[(j/26.9kHz)f+1] [(j/3.92kHz)f+1]
Gain control bandpass
P(f)
35.4 Hz
2.09 kHz
1/0.0357
1/0.0625
Encoded audio output

(c) The zero decibel reference ERMS value for the encoded signal referred to above in (a) is 8.99% modulation of the subcarrier at 0.300 kHz.

Note: This reference results in 0 dB gain through the encoding process at 14.1% modulation using a 0.300 kHz tone, when the ouput bandlimiting filter (see (iv) and (v) following) gain is 0 dB at 0.300 kHz.

(iii) Spectral compression wherein:

(a) The transfer function S(f,b) applied to the audio signal during encoding is:

$$\dot{S}(f,b) = \frac{1 + (jf/F)\ (b+51)/(b+1)}{1 + (jf/F)\ (1+51b)/(b+1)}, \text{ where } b=10^{D/20}$$

F=20.1 kHz; D=decibel rms value and b is the decibel ERMS value of the encoded signal (the result of the encoding process) weighted according to a frequency transfer function Q (f) as follows:

$$Q(f)= \frac{(jf/5.86)^3}{((jf/7.66)^2+(jf/7.31)+1).((jf/26.9)+1).((jf/3.92)+1)}$$

(b) The exponential time weighting period T_2 of the ERMS detector referred to above in (iii-a) is 11.4ms.

(c) The ERMS zero decibel reference for the encoded signal referred to above in (iii-a) is 5.16% modulation of the subcarrier at 8 kHz.

Note: This reference results in +18.4 dB gain through the encoding process at 32.0% modulation using a 8 kHz tone, when the output bandlimiting filter (see (iv) and (v) following) gain is +18.4 dB at 8 kHz.

(iv) Overmodulation protection which functionally follows the functions i, ii, & iii above.

(v) Bandlimiting to appropriately restrict bandwidth which functionally follows the functions i, ii, & iii above.

Television non-program related aural subcarrier standards.

Multiplexing of the aural carrier is subject to the requirements of Section 73.682 (c) of the FCC Rules and Regulations; provided, however, that when the stereophonic and/or second audio program subchannels are transmitted, multiplexing of the aural carrier by non-program related subchannels is subject to the following changes:

(i) The maximum modulation of the aural carrier by the non-program related subcarrier shall comply with the requirements of (D)(a)(1)(vi) of this bulletin.

(ii) When the stereophonic and second audio program subcarriers are transmitted, the instantaneous frequency of the non-program related subcarriers shall have the average value of six and one half times the horizontal scanning frequency with a tolerance of ±500 Hz.

(iii) When only the stereophonic subcarrier is transmitted, the instantaneous frequency of the non-program related subcarrier shall lie between 47 and 120 kHz with a tolerance of ±500 Hz.

TRANSMISSION SYSTEM REQUIREMENTS/ MULTICHANNEL SOUND REQUIREMENTS

Electrical Performance Standards for Stereophonic Operation

The aural transmitter must operate satisfactorily with a frequency deviation of ±73 kHz.It is recommended that the transmitter operate satisfactorily with a frequency deviation of ±100 kHz.

The pilot subcarrier shall be frequency locked to the horizontal scanning frequency of the transmitted video signal.

The requirements of Section 73.687(b)(2) of the FCC Rules and Regulations shall be complied for both the (L+R) main channel and (L−R) subchannel, except for pre-emphasis as specified in Section B)(c) of this bulletin, with the additional requirement that the aural transmitter shall be capable of transmitting a band of frequencies from 50 to 120,000 Hz.

Unless otherwise specified, the transmission system requirements are defined for 75μs pre-emphasis (which is matched to that in the main channel in the case of stereophonic transmission) substituted for encoding. Measurements are made over the band of 50 to 15,000 Hz and employ 75μs de-emphasis in the measuring equipment.

The stereophonic subcarrier, being the second harmonic of the pilot signal, shall cross the time axis with a positive slope simultaneously with each crossing of the time axis by the pilot subcarrier. The pilot subcarrier shall cross the time axis at points located within ± 3 degrees (approximately ± 0.53 microseconds) of the zero crossings of the stereophonic subcarrier.

The unmodulated stereophonic subcarrier shall be suppressed to a level less than 0.25 kHz deviation of the main carrier.

The combined audio frequency harmonics measured at the output of the transmitting system (including the sound encoder), as defined in 73.687 (b) (3) of the FCC Rules and Regulation, at any audio frequency from 50 - 15,000 Hz and at modulating percentages of 25, 50 and 100%, 75μs equivalent modulation shall not exceed the rms values in the following table:

50 to 100 Hz	3.5%
100 to 7,500 Hz	2.5%
7,500 to 15,000 Hz	3.0 %

Harmonics shall be included to 30 kHz.

The ratio of peak main channel deviation to the peak stereophonic subchannel deviation when only a steady state left (or right) signal exists shall nominally be one half for all levels of this signal and for all frequencies from 50 - 15,000 Hz.

The phase and amplitude characteristics of the stereophonic sum modulating signal and the stereophonic difference encoded signal shall be such that the minimum equivalent input separation at 10%, 75μs equivalent modulation is as follows:

(i) 30 dB separation from 100 Hz to 8 kHz.

(ii) Smoothly decreasing separation below 100 Hz, from 30 dB to 26 dB at 50 Hz.

(iii) Smoothly decreasing separation above 8 kHz, from 30 dB to 20 dB at 15 kHz.

Note: it is recommended that the transmission system, excluding encoding, shall meet a 40 dB separation requirement when 75μs pre-emphasis is substituted for sound encoding.

Crosstalk into the main channel caused by a signal in the stereophonic subchannel shall be at least 40 dB below 24 kHz main carrier deviation.

Crosstalk into the main channel caused by a non-stereophonic multiplex signal shall be at least 60 dB below 25 kHz aural carrier deviation.

Crosstalk into the stereophonic subchannel caused by a signal in the main channel shall be at least 40 dB below 50 kHz aural carrier deviation.

Crosstalk into the stereophonic subchannel caused by another multiplex signal shall be at least 60 dB below 50 kHz aural carrier deviation.

The aural transmitting system output frequency modulation noise level in band of 50 - 15,000 Hz (with de-emphasis) must be at least 58 dB below the audio level representing a frequency deviation of ± 25 kHz. The frequency modulaton noise level in the stereophonic subchannel, after demodulation, in the band of 50 - 15,000 Hz (with de-emphasis) must be at least 55 dB below the audio level representing a frequency deviation of ± 50 kHz.

The pilot subcarrier-to-interference ratio, over a bandwidth of 1 kHz centered at the pilot subcarrier, shall be at least 40 dB.

ELECTRICAL PERFORMANCE STANDARDS FOR SECOND PROGRAM OPERATION

The aural transmitter frequency deviation capability must comply with the requirement of Section C(a)(1) of this Bulletin.

The aural transmitter modulation bandwidth capability shall comply with the requirement of Section C(a)(3) of this Bulletin.

The unmodulated subcarrier shall be frequency locked to the fifth harmonic of horizontal line rate. When modulated, the center frequency shall nominally be that of the fifth harmonic of the horizontal line scanning frequency with a tolerance of ± 500 Hz.

Frequency modulation of the subcarrier shall be used.

The subcarrier shall be shut off when the second audio program subchannel is not in use.

The combined audio frequency harmonics measured at the output of the transmitting system (including the encoder) at any audio frequency from 50 - 10,000 Hz and at modulating percentages of 25, 50 and 100% 75μs equivalent modulation shall not exceed the rms values in the following table.

50 to 100 Hz	3.5%
100 to 5,000 Hz	4.0%
5,000 to 10,000 Hz	3.0%

Harmonics shall be included to 20 kHz.

Crosstalk into the SAP subchannel caused by a signal in the main channel and/or in the stereophonic subchannel shall be at least 50 dB below full modulation of the SAP subcarrier (±10 kHz deviation).

The aural transmitting system output frequency modulation noise level after subcarrier demodulation must be at least 50 dB below the level representing full modulation of the SAP subcarrier (±10 kHz deviation).

The aural transmitting system output amplitude modulaton noise level in the band of 63,000-94,000 Hz shall be at least 50 dB below the level representing 100% amplitude modulation.

Electrical Performance Standards for Video Transmission

The requirements of 73.687 (a) (1) and (2) of the FCC Rules and Regulations shall be compiled with; provided, however, that when the station is engaged in stereophonic sound transmission, or when the station transmit stereophonic sound and/or second audio program, subparagraphs (1) and (2) apply, except as modified by the following: A sine wave of 4.5 kHz introduced at the terminals of the transmitter which are normally fed the composite color picture signal shall produce a radiated signal having an amplitude (as measured with a diode on the rf transmission line supplying power to the antenna after the combination of visual and aural power) which is down at least 30 dB with respect to the signal produced by a sine wave of 200 kHz.

In the situation where stereophonic sound and/or a second audio program is transmitted, the following requirements shall be met: the incidental phase modulation of the visual carrier by video signals in the frequency band between 1 and 92 kHz shall be less than three degrees for carrier amplitude below 0.75 of the voltage at synchronizing peaks and less than 5 degrees for carrier amplitudes exceeding 0.75 of the voltage at synchronizing peaks.

Electrical Performance Standards for Sound Encoding

The equivalent input noise of the sound encoder, measured over a 15 kHz bandwidth, shall be more than 70 dB below the 100 Hz, 100% 75μs equivalent modulation level.

The tracking characteristics of the sound encoder shall be such that the minimum equivalent input separation at modulation percentages from 1% to 100% 75 μs equivalent modulation is 26 dB from 100 Hz to 8 kHz.

MODULATION LEVELS

When only a monophonic audio signal is transmitted, the modulation of the aural carrier shall not exceed 25 kHz deviation on peaks of frequent recurrence, unless some other peak modulation level is specified.

For stations transmitting more than one audio program channel the maximum modulation levels must meet the following limitations:

(1) TV stations transmitting stereophonic sound signals must limit the modulation of the aural carrier by the stereophonic sum modulating signal to 25 kHz deviation on peaks of frequent recurrence.

(2) TV stations transmitting stereophonic sound signals must limit the modulation of the aural carrier by the sum of stereophonic sum modulating signal and stereophonic difference encoded signal to 50 kHz deviation on peaks of frequent recurrence.

(3) The modulation of the aural carrier by the stereophonic pilot signal shall be 5.0 kHz deviation with a tolerance of ± 0.5 kHz.

(4) TV stations transmitting a second audio program must limit the modulation of the aural carrier by the SAP subcarrier to 15 kHz deviation.

(5) TV stations transmitting multiplex signals on the aural carrier for non-program related purposes must limit the modulation of the aural carrier by the arithmetic sum of all subcarriers, other than the stereophonic and second audio program, to 3 kHz deviation.

(6) Total modulation of the aural carrier by multichannel sound shall not exceed 73 kHz.

DEFINITIONS

Compandoring: A noise reduction process used in the stereophonic subchannel and the second audio program subchannel consisting of encoding (compression) before transmission and decoding (expansion) after reception.

Composite stereophonic baseband signal: The stereophonic sum modulating signal, the stereophonic difference encoded signal and the pilot subcarrier.

Crosstalk: An undesired signal occurring in one channel caused by an electrical signal in other channels.

Decibel ERMS value: The exponentially time-weighted root mean square (ERMS) value converted to dB as follows:

$$\text{decibel ERMS value} = 20 \log_{10} \frac{(\text{ERMS value})}{(\text{Reference})}$$

Where Reference is the 0dB ERMS value.

Encoding: See compandoring

Equivalent input separation: A method of specifying the stereophonic separation by referring variations from ideal at the output back to the input. To accomplish this, an input signal which causes a non-ideal output is varied by degrading input separation until the output conforms to the ideal. The amount of input separation degradation required is the equivalent input separation.

Equivalent modulation: See equivalent input tracking.

Equivalent input tracking: A method of specifying the tracking ability of the encoding process by referring variations from ideal at the output back to the input of the encoder. To accomplish this, an input signal which causes a non-ideal output is varied until the output conforms to the ideal. The amount of input variation required is the equivalent input tracking.

Expoentially time-weighted root mean square (ERMS) value: The ERMS value of a waveform is obtained from the following formula:

$$\text{ERMS value} = \sqrt{\frac{1}{T} \int_{-\infty}^{t} S^2(u)e^{-(t-u)/T}du}$$

Where S(u) is the waveform in question, a function of time, T is the exponential time-weighing period, and t is the time at which the ERMS value is computed.

Incidental carrier phase modulation (ICPM): Angle modulation of the visual carrier by video signal components which, when detected in TV receiver intercarrier circuits, cause an audio interference known as intercarrier buzz.

Left (or right) audio signal: The electrical output of a microphone or combination of microphones placed so as to convey the intensity, time, and location of sounds originating predominately to the listener's left (L) (or right (R)) of the center of the performing area.

Left (or right) stereophonic channel: The transmission path for the left (or right) audio signal.

Main channel: The band of frequencies from 50 to 15,000 Hz which frequency modulates the aural carrier.

Multichannel sound: Multiplex transmission on the TV aural carrier.

Multiplex transmission: The simultaneous transmission of the TV program main channel audio signal and one or more subchannel signals. The subchannels include a stereophonic subchannel, a second audio program subchannel, a non-program related subchannel and a pilot subcarrier.

Non-program related subchannel: The subchannel for the multiplex transmission of a frequency-modulated subcarrier for telemetry or other purposes.

Pilot subcarrier: A subcarrier serving as the control signal for use in the reception of TV stereophonic sound broadcasts.

Second audio program (SAP) broadcast: The multiplex transmission of a second audio program utilizing the second audio program subchannel.

Second audio program (SAP) subchannel: The channel containing the frequency-modulated second audio program subcarrier.

Second program audio signal: The monophone audio signal delivered to the SAP encoder.

Second program encoded signal: The second program audio signal after encoding.

75 microsecond (75μs) equivalent modulation: The audio signal level prior to encoding that results in a stated percentage modulation when the encoding process is replaced by 75μs pre-emphasis.

Spectral compression: A process wherein variations in spectral content of an audio signal are reduced by varying a frequency filtering function applied to the signal in response to variations in the spectral content of the signal.

Stereophonic difference audio signal: The left audio signal minus the right audio signal (L-R).

Stereophonic difference encoded signal: The stereophonic difference audio signal after encoding.

Stereophonic separation: The ratio of the electrical signal caused in the right (or left) stereophonic channel to the electrical signal caused in the left (or right) stereophonic channel by the transmission of only a right (or left) signal.

Stereophonic sum audio signal: The left audio signal plus the right audio signal PL+R).

Stereophonic sum channel compensation: A process wherein the phase and amplitude response resulting from bandlimiting in the process of encoding the stereophonic difference audio signal (which, if uncompensated, would detrimentally affect stereo separation) is compensated by an identical phase and amplitude response applied to the stereophonic sum audio signal.

Stereophonic sum modulating signal: The stereophonic sum audio signal after compensation, pre-emphasis and other processing.

Stereophonic subcarrier: A subcarrier having a frequency which is the second harmonic of the pilot subcarrier frequency and which is employed in TV stereophonic sound broadcasting.

Stereophonic subchannel: The subchannel containing the stereophonic subcarrier and its associated sidebands.

Wideband amplitude compression: A process where in the dynamic range of an audio signal is compressed by simultaneously varying the gain of all audio frequencies equally.

Index

OTHER POPULAR TAB BOOKS OF INTEREST

Transducer Fundamentals, with Projects (No. 1693—$14.50 paper; $19.95 hard)

Second Book of Easy-to-Build Electronic Projects (No. 1679—$13.50 paper; $17.95 hard)

Practical Microwave Oven Repair (No. 1667—$13.50 paper; $19.95 hard)

CMOS/TTL—A User's Guide with Projects (No. 1650—$13.50 paper; $19.95 hard)

Satellite Communications (No. 1632—$11.50 paper; $16.95 hard)

Build Your Own Laser, Phaser, Ion Ray Gun and Other Working Space-Age Projects (No. 1604—$15.50 paper; $24.95 hard)

Principles and Practice of Digital ICs and LEDs (No. 1577—$13.50 paper; $19.95 hard)

Understanding Electronics—2nd Edition (No. 1553—$9.95 paper; $15.95 hard)

Electronic Databook—3rd Edition (No. 1538—$17.50 paper; $24.95 hard)

Beginner's Guide to Reading Schematics (No. 1536—$9.25 paper; $14.95 hard)

Concepts of Digital Electronics (No. 1531—$11.50 paper; $17.95 hard)

Beginner's Guide to Electricity and Electrical Phenomena (No. 1507—$10.25 paper; $15.95 hard)

750 Practical Electronic Circuits (No. 1499—$14.95 paper; $21.95 hard)

Exploring Electricity and Electronics with Projects (No. 1497—$9.95 paper; $15.95 hard)

Video Electronics Technology (No. 1474—$11.50 paper; $16.95 hard)

Towers' International Transistor Selector—3rd Edition (No. 1416—$19.95 vinyl)

The Illustrated Dictionary of Electronics—2nd Edition (No. 1366—$18.95 paper; $26.95 hard)

49 Easy-To-Build Electronic Projects (No. 1337—$6.25 paper; $10.95 hard)

The Master Handbook of Telephones (No. 1316—$12.50 paper; $16.95 hard)

Giant Handbook of 222 Weekend Electronics Projects (No. 1265—$14.95 paper)

Introducing Cellular Communications: The New Mobile Telephone System (No. 1682—$9.95 paper; $14.95 hard)

The Fiberoptics and Laser Handbook (No. 1671—$15.50 paper; $21.95 hard)

Power Supplies, Switching Regulators, Inverters and Converters (No. 1665—$15.50 paper; $21.95 hard)

Using Integrated Circuit Logic Devices (No. 1645—$15.50 paper; $21.95 hard)

Basic Transistor Course—2nd Edition (No. 1605—$13.50 paper; $19.95 hard)

The GIANT Book of Easy-to-Build Electronic Projects (No. 1599—$13.95 paper; $21.95 hard)

Music Synthesizers: A Manual of Design and Construction (No. 1565—$12.50 paper; $16.95 hard)

How to Design Circuits Using Semiconductors (No. 1543—$11.50 paper; $17.95 hard)

All About Telephones—2nd Edition (No. 1537—$11.50 paper; $16.95 hard)

The Complete Book of Oscilloscopes (No. 1532—$11.50 paper; $17,95 hard)

All About Home Satellite Television (No. 1519—$13.50 paper; $19.95 hard)

Maintaining and Repairing Videocassette Recorders (No. 1503—$15.50 paper; $21.95 hard)

The Build-It Book of Electronic Projects (No. 1498—$10.25 paper; $18.95 hard)

Video Cassette Recorders: Buying, Using and Maintaining (No. 1490—$8.25 paper; $14.95 hard)

The Beginner's Book of Electronic Music (No. 1438—$12.95 paper; $18.95 hard)

Build a Personal Earth Station for Worldwide Satellite TV Reception (No. 1409—$10.25 paper; $15.95 hard)

Basic Electronics Theory—with projects and experiments (No. 1338—$15.50 paper; $19.95 hard)

Electric Motor Test & Repair—3rd Edition (No. 1321 $7.25 paper; $13.95 hard)

The GIANT Handbook of Electronic Circuits (No. 1300—$19.95 paper)

Digital Electronics Troubleshooting (No. 1250—$12.50 paper)

TAB **TAB BOOKS Inc.**

Blue Ridge Summit, Pa. 17214

Send for FREE TAB Catalog describing over 750 current titles in print.